高等院校土建学科双语教材（中英文对照）
◆ 城市规划专业 ◆
BASICS

城市街区
URBAN BUILDING BLOCKS

[德] 托尔斯滕·别克林　编著
　　　迈克尔·彼得莱克
张路峰　译

中国建筑工业出版社

著作权合同登记图字:01-2009-7707号

图书在版编目(CIP)数据

城市街区/(德)别克林等编著;张路峰译. —北京:中国建筑工业出版社,2011.4
高等院校土建学科双语教材(中英文对照)◆ 城市规划专业 ◆
ISBN 978-7-112-13029-0

Ⅰ.①城… Ⅱ.①别…②张… Ⅲ.①城市规划-汉、英 Ⅳ.①TU984

中国版本图书馆 CIP 数据核字(2011)第 043467 号

Basics: Urban Building Blocks / Thorsten Bürklin, Michael Peterek
Copyright © 2008 Birkhäuser Verlag AG (Verlag für Architektur), P. O. Box 133, 4010 Basel, Switzerland
Chinese Translation Copyright © 2011 China Architecture & Building Press
All rights reserved.
本书经 Birkhäuser Verlag AG 出版社授权我社翻译出版

责任编辑:孙　炼
责任设计:陈　旭
责任校对:陈晶晶　马　赛

高等院校土建学科双语教材(中英文对照)
◆ 城市规划专业 ◆

城市街区

[德] 托尔斯滕·别克林　编著
　　 迈克尔·彼得莱克

张路峰　译

*

中国建筑工业出版社出版、发行(北京西郊百万庄)
各地新华书店、建筑书店经销
北京嘉泰利德公司制版
北京云浩印刷有限责任公司印刷

*

开本:880×1230 毫米　1/32　印张:4　字数:160 千字
2011 年 6 月第一版　　2011 年 6 月第一次印刷
定价:**18.00** 元
ISBN 978-7-112-13029-0
(20433)

版权所有　翻印必究
如有印装质量问题,可寄本社退换
(邮政编码　100037)

中文部分目录

\\ 序　7

\\ 导言：从单栋房屋到城市街区　75

\\ 联排式　76
　　\\ 形态与空间结构　76
　　\\ 城市空间构成　78
　　\\ 功能、朝向和连接方式　79
　　\\ 历史案例　81

\\ 街坊式　84
　　\\ 形态与空间结构　84
　　\\ 城市空间构成　85
　　\\ 功能、朝向与连接方式　86
　　\\ 历史案例　88

\\ 庭院式（反转的街坊式）　93
　　\\ 形态与空间结构　93
　　\\ 城市空间构成　94
　　\\ 功能、朝向和连接方式　94
　　\\ 历史案例　95

\\ 街廊式　98
　　\\ 形态与空间结构　98
　　\\ 城市空间构成　98
　　\\ 功能、朝向和连接方式　99
　　\\ 历史案例　100

\\ 行列式　102
　　\\ 形态与空间结构　102
　　\\ 城市空间构成　104
　　\\ 功能、朝向和连接方式　104
　　\\ 历史案例　106

\\ 独栋式　108
　　\\ 形态与空间结构　108
　　\\ 城市空间构成　110
　　\\ 功能、朝向和连接方式　111
　　\\ 历史案例　113

\\ 组团式　116
　　\\ 形态与空间结构　116
　　\\ 城市空间构成　117
　　\\ 功能、朝向和连接方式　118
　　\\ 历史案例　118

\\ 棚厦式　121
　　\\ 形态与空间结构　122
　　\\ 城市空间构成　122
　　\\ 功能、朝向与连接方式　122
　　\\ 历史案例　123

\\ 结语：从城市街区到城市结构　125

\\ 附录　127
　　\\ 参考文献　127
　　\\ 作者简介　128

CONTENTS

\\Foreword _9

\\Introduction: From individual building to building block of the city _11

\\The row _13
    \\Form and spatial structure _13
    \\Formation of urban space _15
    \\Functions, orientation and access _17
    \\Historical examples _19

\\The city block _22
    \\Form and spatial structure _22
    \\Formation of urban space _24
    \\Functions, orientation and access _25
    \\Historical examples _27

\\The courtyard (inverse block) _33
    \\Form and spatial structure _33
    \\Formation of urban space _34
    \\Functions, orientation and access _35
    \\Historical examples _36

\\The arcade _39
    \\Form and spatial structure _39
    \\Formation of urban space _40
    \\Functions, orientation and access _41
    \\Historical examples _42

\\The ribbon _44
    \\Form and spatial structure _44
    \\Formation of urban space _46
    \\Functions, orientation and access _47
    \\Historical examples _48

\\The solitaire _51
    \\Form and spatial structure _51
    \\Formation of urban space _51
    \\Functions, orientation and access _54
    \\Historical examples _56

\\The group _61
  \\Form and spatial structure _61
  \\Formation of urban space _62
  \\Functions, orientation and access _62
  \\Historical examples _63

\\The shed _66
  \\Form and spatial structure _67
  \\Formation of urban space _67
  \\Functions, orientation and access _68
  \\Historical examples _69

\\In conclusion: From building block of the city to urban structure _71

\\Appendix _73
  \\Literature _73
  \\The authors _74

序

　　建筑总是作为自然环境或人工环境的一部分而不是孤立存在的。该环境由历史、文化及景观的关联网络而构成，而在当今世界，设计的实施和建筑的落成基本上都是在城市的环境之中。只要城市的发展势头足够强大，就能够保障大规模的城市项目在图板上被描绘，并且被完整地实现，然而城市规划师和建筑师所面对的城市，却是一个不断演变的、经受各种不同理念影响的城市。尽管存在这种复杂性，我们发现"城市街区"对于城市环境的形成和影响是最大的。

　　本套丛书旨在为学生提供一个实用的、指导性的城市设计基础性介绍。本书描述了典型的城市街区类型及其特点，使学生对城市的基本构成和设计有一个初步的了解。该书阐释了联排式、街坊式、庭院式、街廊式、行列式、独栋式、组团式和棚厦式等类型，并重点分析了各种类型的构成原理、功能使用以及文化历史背景。在城市规划分析和设计工作中，无论其目的是创造新的城市环境，还是对原有的城市结构进行更新，或者在城市的肌理中去设计一栋单体建筑，对城市设计中所使用的要素的把握都是非常重要的。

贝尔特·比勒费尔德

套书主编

FOREWORD

Buildings do not emerge in isolation but as part of a natural or built environment. This environment can consist of a web of historical, cultural and landscape relations, but in today's world, designs are implemented and architecture emerges primarily in an urban context. A large-scale urban project can be designed on the drawing board and built as a coherent unit if the city's growth potential warrants this approach, but urban planners and architects generally create designs for cities that have evolved over generations and are the products of different influences and ideologies. Despite this complexity, we find recurring "building blocks" that shape and influence the urban environment.

The field of "urban planning" in the *Basics* series provides students with a practical, instructive introduction to the foundations of urban design. The present book describes typical urban building blocks and their features in order to give students a basic understanding of the fundamental structure and design of cities. It elucidates the row, block, courtyard, arcade, ribbon, solitaire, group and shed, focusing on structural principles, functional possibilities and cultural historical backgrounds. Knowledge of the elements used to design cities is important for the analytical and creative work in urban planning, regardless of whether the goal is to create new urban environments, add to or renew existing urban structures, or to design individual buildings in an urban area.

Bert Bielefeld
Editor

INTRODUCTION:
FROM INDIVIDUAL BUILDING TO BUILDING BLOCK OF THE CITY

The city is more than the sum of its individual buildings. It is also more than "large-scale architecture". In its neighbourhoods and quarters– the arenas of our day-to-day lives – it is made up of built structural elements that mediate between the scale of the individual architectural objects and that of larger units such as neighbourhoods or even entire urban districts. These elements thus mediate between the individuality (and privacy) of a house and plot of land and the collective (and public) sphere of a more comprehensive urban environment.

These structural elements can also be termed "building blocks of the city". They can appear as different forms and geometries in the urban layout: rows, blocks, courtyards, arcades, ribbons, solitaires, groups and "sheds". Naturally, diverse combinations can be imagined and are already part of urban reality. By virtue of their special form and unique combinations, they influence the way we live together by promoting certain functions and lifestyles and hindering others. Knowledge of these building blocks is therefore an essential aspect of the craft of urban design. Urban planners and architects must grapple with them in order to evaluate the effects that their designs will have. It is only by understanding these urban elements – which differ greatly in terms of form, function, size and significance – that they can responsibly design cities.

The following chapters present these building blocks from different perspectives, focusing on their spatial structure and design, functional objectives, integration into the urban environment, the associated differentiation of private and public areas, and the conditions under which they emerge. Where relevant, the chapters also touch on the way these basic structural elements of the city have changed over the years. Individual observations are illustrated using historical and modern examples.

Each building block is discussed in relation to the following four points:

_ Form and spatial structure (physical description of the urban element)
_ Formation of urban space (the impact of the "building block" and the significance it has for its surroundings and for urban space)

_ Functions, orientation and access
_ Historical examples

Of course the distinctions between these building blocks are not always as clear in architectural and urban reality as the thematic structure of this book might suggest. There are a great many overlaps, borderline cases and "hybrids" that do not fall into a clearly definable category.

Even so, it is important that students first study the building blocks of the city in their purest form so that they can use this knowledge to analyse the different combinations and hybrid forms found in cities and take them into account sufficiently in their designs. With this in mind, this book aims to provide formal, functional and organizational information and knowledge concerning the individual building blocks of the city.

THE ROW

The row is one of the oldest and most important structural elements in cities and settlements. It joins together individual plots of land and buildings along a straight, angular or curved line, formed and accessed by the street. It creates a broader urban planning context that extends beyond the individual building. The basic spatial structure of large areas of our cities and villages consists of rows.

FORM AND SPATIAL STRUCTURE

Relation to the street

Constitutive of the row is that the buildings' entrances and access paths are oriented toward the street, which defines the row spatially and functionally. > Fig. 1

Development forms

In addition to the principle of linear addition, rows can have entirely different development forms. They can be open or closed (i.e. terraces), and have one or two sides. In open rows of single-family or semi-detached homes, there is open space surrounding the buildings. Whereas single-family homes have open space on all four sides, each semi-detached house is joined to its twin on one side. In terraces, there are no gaps between the buildings, which form a continuous visible spatial edge.

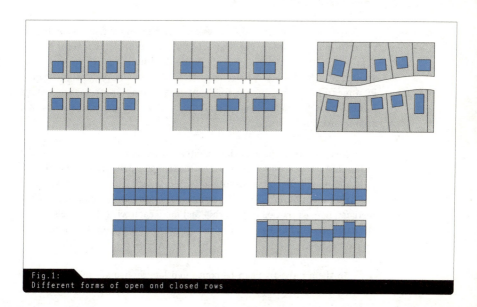

Fig.1:
Different forms of open and closed rows

Fig. 2:
One-sided row on a slope in Bath in the south of England

In a one-sided row (which can be either open or closed), only one side of the street is developed; a two-sided row has buildings on both sides. The buildings on opposite sides of a two-sided row need not necessarily be identical. The two sides of the row are formally independent of each other. The row, particularly in its open form, can be excellently adapted to dynamic site topographies. › Fig. 2

Diversity in unity

The row is a highly flexible urban building block and allows for diverse formal principles. The appearance, three-dimensional form (width, depth, height) and functions of the individual buildings in a row can be similar or even absolutely identical. › Fig. 3 They can also differ significantly in appearance, with highly differentiated, irregular or heterogeneous forms. › Fig. 4 This means that each individual building can have a distinct appearance and identity.

Even so, rows in an urban context are often arranged so that the individual parts are in harmony. The reasons are primarily economic: the repeated use of the same prototype facilitates the quick and inexpensive construction of buildings and apartments. Single-family, semi-detached or terraced houses that emerge in this fashion – each with the same size and the same internal and external form – leave their mark on the character of an entire area.

In theory, the row can be continued lengthwise ad infinitum. However, infrastructure capacity and long distances place limitations on row

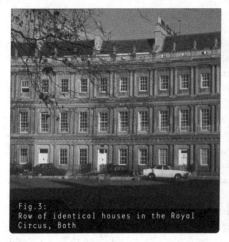

Fig.3:
Row of identical houses in the Royal Circus, Bath

Fig.4:
Row of individual buildings on Amsterdam harbour

length since at some point site development becomes uneconomical. Furthermore, long rows make it impossible to develop the entire depth of the plots behind the buildings as building land. Diagonal streets must be introduced to break up the rows and create more economical urban units in the form of blocks. › Chapter The city block

Economic efficiency is another reason why one-sided rows are usually an exception in urban planning. If a row has two sides, twice as many buildings can be linked to the urban infrastructure with the same expenditure of money and effort. One-sided rows are generally confined to special areas such as the edge of housing estates or the area along parks and rivers, where there is a need for particularly high-quality residential and working space.

FORMATION OF URBAN SPACE

Through their direct link to the street, rows form clear and distinguishable urban spaces. At the same time, they are integrated into the infrastructure network of the entire city, and thus become part of a cohesive urban spatial web. Rows can flexibly fill gaps and intermediate spaces in this urban structure, and they can also be easily connected to other urban elements such as city blocks, ribbons or solitaires.

Front and back

Due to their orientation toward the street, rows are characterized by a clear spatial demarcation between front and rear. This distinction is reflected not only in the different functions (public use on the street side, and private or collective use in the gardens or courtyards), but also in different architectural designs. On the street side, the buildings usually have a more restrained and stately design, while in the rear they may be less regimented and more influenced by significant processes of individual appropriation and alteration (e.g. extensions and remodelling to create terraces, pergolas, winter gardens and roof space).

Transition to street space

The design of the front area facing the street is of great importance for the formation of urban space. This is especially true of the transition from the private space of the building to the public space of the city. Depending on location, orientation, topography, type of building and other factors, this front transitional area can be designed in a variety of ways. In densely developed historical cities, it was once common practice to construct buildings directly on public streets in order to save space. Property boundaries and building lines coincided. This continues to be a suitable solution today if the ground floor of a building is used for commercial purposes, but it is not effective for residential uses because pedestrians are able to look into private spaces if the ground floor is on the same level as the street. In the residential neighbourhoods of most modern cities, architects therefore usually create a deep buffer zone between the private area of the building and public space. This can take the form of a planted strip of land, a front garden or a private front courtyard that provides space for important supplementary functions next to the house (parking spaces, carports, storage space for bikes, rubbish bins, places to relax, sitting areas, terraces etc.). Even a row of trees can serve as a transitional area. By slightly raising the ground floor level, architects can keep passers-by from looking into the house and also provide residents with an attractive view of street space from inside the apartment. › Fig. 5

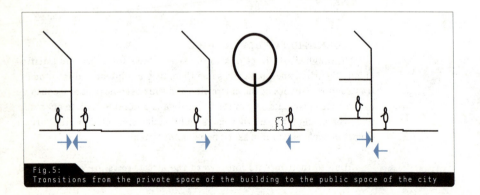

Fig.5:
Transitions from the private space of the building to the public space of the city

FUNCTIONS, ORIENTATION AND ACCESS

Combined uses

Because of its direct link to the network of urban streets and city infrastructure, the row can perform the full gamut of an urban building block's functions. Even today, we find vivid examples of mixed uses in the streets that make up historical city centres and in organically evolved neighbourhoods. The ground floors in row developments are especially well suited to uses beyond mere housing (shops, restaurants and small businesses). If required, these functions can be extended to the rear of the property by means of extensions or additional auxiliary buildings – common practice in closed rows that otherwise offer little space. Entrance drives through the front buildings can be helpful since they provide access to the rear area of the plot even for cars. Yet they subject the quiet, private area protected from the street to additional disturbances from the cars driving in, noise and greater public accessibility.

That said, the row has developed into a primarily monofunctional urban element that is mostly used for housing. This is the result of modern urban planning concepts that, in line with the Athens Charter, separate urban functions into housing, work and transport.

Dense terraced housing

Rows formed by closed groups of terraced houses became popular in the 20th century, particularly because they made economical use of the limited land available for urban expansion. Due to its capacity to encompass a large number of subdivisions, the row allows for individual housing solutions on private plots that are characterized by a high degree of development density, compared to the open rows of single-family or semi-detached homes found elsewhere. In addition to the economic efficiency of standardized building elements, the row offers the advantage of saved space and costs since the individual homes can be built on very narrow pieces of property.

\\ Note:
The Athens Charter was passed in 1933 at the fourth meeting of the Congrès Internationaux d'Architecture Moderne (International Congress of Modern Architecture). The conference was held in July and August 1933 on board the "Patris" sailing from Marseilles to Athens. Le Corbusier was most important initiator and main author of the Charter.

\\ Note:
The width of the plots used for terraced housing usually measures 5.5 to 6.5 m. In some cases, these plots can be as narrow as 4.5 m. If they have a depth of 25 to 30 m, the total surface area is 150 to 180 m^2. By contrast, semi-detached and single-family homes usually require plots of 300 to 400 m^2 or more.

Orientation

Since rows are built along streets, their orientation in relation to the points of the compass and the sun depends on the course of the street. As a consequence, natural and artificial lighting conditions for the houses and apartments may vary greatly with property orientation. Any such locational disadvantages must be compensated for by adequate floor plan design (e.g. floor plans that use both sides of the building), since rows, particularly closed ones, receive sunlight only at certain times of the day.

One special form of terraced housing involves stacking one row on top of the other. This usually involves two duplexes, with the upper unit accessed via an outdoor corridor. This solution can be used to create a high-density urban environment that has the same residential quality and atmosphere offered by separately accessed individual homes. > Fig. 6

Townhouses put to mixed uses

Recently, the term "townhouse" has come to describe buildings in a row (or on a block) that combine the functions of housing and work in a densely developed inner-city area. > Fig. 7 Townhouses are at least three storeys tall and sometimes four. They offer sufficient space for a supplementary commercial use (e.g. a store or an office on the ground floor), and they might also have a (separate) granny flat, a private courtyard or a garden. Some feature additional outdoor space such as a roof terrace. Townhouses are characterized by great architectural diversity and individuality.

\\ Note:
Rows facing east or west receive sufficient sunlight in the evening and the morning, but no direct midday sun. In the summer, the sunlight can heat up the rooms and make it necessary to install sunshades on the buildings. Rows oriented north or south in the northern hemisphere benefit from sunlight coming from the south, which can create pleasant indoor temperatures particularly during the winter months. This orientation offers the added advantage of energy savings (passive solar energy use). However, it has the disadvantage that any rooms on the northern side of the building receive no sunlight at all during the day (with the exception of late afternoon sun at the peak of summer). This is why you should never design, say, a children's room with this orientation. Nevertheless, the northern side of a building is often better suited for studios and certain work spaces since people can work undisturbed if lighting conditions remain constant. It should be noted that such references to north-south orientations in this book should be reversed when considering buildings in the southern hemisphere.

Fig.6:
Stacked rows of duplexes in the densely developed Margess Road estate in London

Fig.7:
Rows of townhouses in the centre of Karlsruhe

HISTORICAL EXAMPLES

The creation of rows made up of similar plots and buildings played an important role in the cities of Antiquity, including the newly founded colonies of ancient Greece. The reason for their popularity is that they offered a simple and rational principle for dividing up urban land. An additional advantage is that they made it possible to treat all residents equally (the same conditions and the same use for all).

Medieval townhouses

The cities of the Middle Ages were also based on rows of plots and buildings. Although these buildings were identical in terms of typology, their architectural details were often quite different. The craftsman's or merchant's house, with its mixed uses, formed the basic unit of the city. In the front it meshed with the public space of the city – with its alleys, streets and squares. In the rear, it overlooked a completely private area consisting of courtyards and gardens that were almost invisible to the outer world. If the urban space was densely developed and more space was required, additional buildings could be built at the rear. Typical examples can be found in numerous medieval towns such as Gdansk, Lübeck and Amsterdam. Cities like these have managed to retain their urban atmosphere and quality of life up until today. › Fig. 8

Garden cities

Historical cities and, in particular, 19th-century industrial cities were criticized for being too densely developed and having cramped living conditions. In response to this, the garden city movement, which

Fig.8:
Rows of medieval merchants' houses in the old part of Gdansk

originated in England in the early 20th century, sought to create new residential estates and urban expansion projects that were modelled on open and closed rows. The urban planning and spatial objectives hinged on creating a diverse, open development structure and were linked to the intention to implement general reforms – social, economic, health and hygiene concerns. Urban planners were particularly keen to provide the poorly housed working class with dwellings in green areas. Deep gardens behind the terraced homes could be used agriculturally to meet the families' needs. › Fig. 9

Modern estates

These concepts also played a role in the design of terraced housing estates by the *Neues Bauen* movement in the early 20th century – as can be seen in the residential estates of Bruno Taut and Martin Wagner in Berlin

\\ Note:
The central work of the garden city movement is *Tomorrow: A Peaceful Path to Real Reform* by Ebenezer Howard, published in 1898.
Howard wanted the garden city to combine the advantages of both the city and the country. Since these cities were supposed to be places to live and work, they included both industrial and cultural facilities. According to Howard's concept, six largely autonomous cities (each with 32,000 residents) were grouped around a central urban area (with 58,000 residents). However, only a few of these autonomous garden cities were ever built. The first was Letchworth, founded north of London in 1904. Most were planned as garden suburbs on the peripheries of existing cities, upon which they remained functionally and economically dependent.

Fig.9:
Terraced houses in the Karlsruhe garden city suburb

Fig.10:
Terraced houses in Römerstadt, Frankfurt

and the projects built in Frankfurt am Main under Ernst May, director of the municipal planning office. The Römerstadt estate (1927–1928) is based on the principle of rowing together standardized, economically laid-out homes, each with its own garden. This arrangement creates a concise urban street space with a pleasant atmosphere. > Fig. 10

The current renaissance of townhouses shows that the row continues to have great appeal. It is an urban housing form with tremendous individuality that supports a variety of lifestyles. At the same time, the uncontrolled development of the areas surrounding our cities – a result of the construction of single-family homes that consume an inordinate amount of land – underscores the urgent need for sustainable housing models like the row that require less space.

\\ Note:
The *Neues Bauen* architectural movement emerged around the Bauhaus, a school that was opened in Weimar in 1919 with workshops for crafts, architecture and the visual arts.
The movement's primary goal was to transcend historicism and create rational architecture that made use of industrial production methods.

THE CITY BLOCK

Like the row, the city block (or block of buildings) is one of the oldest and most important elements of urban design. From Antiquity onward, it has exerted a major influence on the structure of European cities. However, in the early 20th century, urban planners argued that it created inequitable living conditions, and it was not until the end of the century that its positive qualities as an urban element were rediscovered.

FORM AND SPATIAL STRUCTURE

Outside and inside

The block consists of a group of plots – or, in special cases, of a single property – and it is surrounded and accessed by streets on all sides. The front facades of the buildings forming the block are oriented toward the street, creating a clear distinction between the block's interior and exterior space and a strong architectural orientation toward a front public area and a rear private realm. The block's interior may be left open or partially or fully covered with buildings. It may be used for gardens, courtyards, open areas, garages, storage spaces, ancillary buildings and so on.

Block geometries

Blocks can have a wide range of geometries. They can be triangular, rectangular, square, polygonal, oval, semicircular or even circular. The decisive factor is that, on all sides, they are accessed by and oriented toward the outside area. Even so, their basic geometric shape leads to different frameworks for architectural and urban design (e.g. sharp corners), the design and quality of the interior areas, and lighting conditions in apartments.

Blocks of buildings can be closed on all sides, or the edges can be interrupted and contain gaps. An open city block consists of short rows of terraced, semi-detached or single-family homes, but they must be situated so close to each other that they do not mar the impression of a block and appear to be solitaires. › Fig. 11

Designing corners

A special challenge–and not only from an architectural perspective–lies in designing the corners of a block. For one thing, corners have a particularly favourable position (for shops, restaurants and other commercial facilities) because they can be accessed from both sides. For another, they are a critical point with a number of disadvantages: the rear (property) area is very small or there may be no rear space at all. Corners are unsuitable for private uses or expansion, and the narrow rear facade

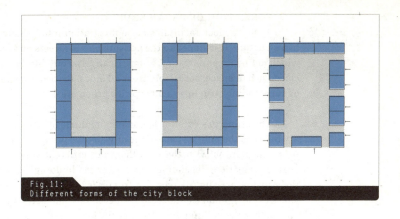

Fig. 11:
Different forms of the city block

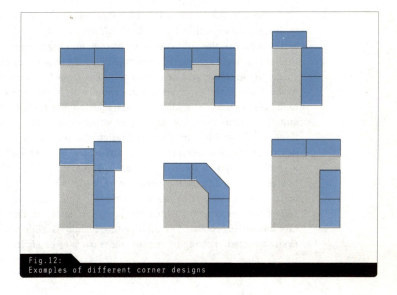

Fig. 12:
Examples of different corner designs

may receive inadequate natural lighting, depending on the building's orientation.

The corners of city blocks can be designed with gaps so that the corner buildings receive adequate light. They can also be completely removed or "bevelled". Another possibility is to create especially wide or narrow corner buildings. › Fig. 12

FORMATION OF URBAN SPACE

Integration into the city

The city block facilitates close integration into the surrounding urban structure. It is linked to the network of city streets and building lines, which define it spatially and geometrically. The city block is a continuous closed urban space, accessible from all sides, that ensures the continuity of surrounding structures and exterior urban areas.

The property's exterior boundary simultaneously defines the boundary between the public space of the city and the private space of the buildings and plot. As with the row, there are different ways to design this spatial transition, depending on whether the buildings stand directly on the street or are slightly recessed from it (e.g. through a front garden), on whether the ground floors serve residential or commercial purposes; and on whether the building has a basement floor. > Chapter The row

Front and rear

The different manner in which the front and rear sides are treated in designs reflects the clear spatial differentiation between the exterior (with its link to the public area of the city) and the interior (with its link to shared private space). This treatment covers not only the design of open space, but architectural design as well. The front facades with their link to the street, visible to everyone, are usually designed to meet a relatively high creative standard. Materials are selected which have an impressive, stately character, and the horizontal and vertical structure, the proportions and the architectural ornamentation of windows are subject to a high degree of creative discipline. By contrast, the rear, which is visible and accessible only to a limited degree to the general public and neighbours, is often designed to meet practical needs. Windows do not adhere to geometric organizational principles as rigidly, and their size and positions reflect the purposes for which they are used (kitchens, bathrooms, ancillary and sitting rooms). The architecture is more flexible and can be more easily adapted to changing requirements (such as extensions, remodelling

> \\Note:
> In 20th-century city blocks, which tend to have a more uniform design, the interior and exterior facades were often treated similarly due to a new conception of architecture and public space. The same is true of the otherwise different spatial characters of the inner and outer areas.

projects and conversions). All told, the city block is a spatial system that is extremely complex and flexible, and lends itself particularly well to integrating diverse, differentiated modes of behaviour, activities and forms of appropriation.

High density

By virtue of its rational and economic use of urban land, the city block allows a relatively high degree of urban density. This can be regarded as an important environmental and economic advantage given the current discussion of the increased use of land in the regions surrounding our cities.

FUNCTIONS, ORIENTATION AND ACCESS

Mixed uses

The city block is well suited to diverse functions and combined uses because of its direct integration into the broader spatial system, streets and squares of the city. Although the ground floors of buildings on a block are close to the street and lack privacy, they have proved an excellent location for shops, small businesses and restaurants over the centuries.

The flexible rear area of the city block can provide space for numerous activities and uses that find their architectural expression in supplementary buildings. In the Middle Ages and the *Gründerzeit*, there were often workshops inside blocks, and work and housing were closely intertwined. One also sees examples of entire factory buildings located inside a city block. › Fig. 13 Where required, the rear courtyards were accessed by entrance drives leading through the front buildings.

In the early 20th century, there was a move to banish the disruptive businesses from inside city blocks due to their noise and pollution. This occurred in connection with the Athens Charter, which recommended creating a clear separation between urban functions such as housing, work, recreation and transport. › Chapter The row Ever since, blocks have primarily been used for housing, and the internal area accommodates private and collective playgrounds, open spaces, gardens and planted areas.

It was not until the 1970s that balanced combinations of non-disruptive functions were once again introduced into the city. The transition from industrial to service society has changed workplaces, the possible disturbances they cause, and the ease with which they can be integrated into the surrounding residential area of a city quarter. In most cases, combined uses are not a problem, and indeed can create a special quality. The city block continues to offer excellent conditions for such combinations, even if not all blocks in a city or a neighbourhood have the same degree of density as regards commercial uses. This density is usually greatest

Fig.13:
Different ways that block interiors are used

along main traffic routes, with nearby ground floors usually being used for residential purposes.

Building depth and orientation

Since the edges of a city block follow the course of the street, there are limitations on building orientation. In apartments oriented to both sides of the building, auxiliary rooms can be located in the interior area that is not illuminated by natural light (this is assuming that the average depth of east-west buildings is 11 to 13 m). Floor plans with a predominantly north-south orientation should be wider and have a shallower depth amounting to only about 9 to 11 m. This allows more effective use of the southern facade, which receives direct sunlight.

Aside from sunlight, other important factors for orienting apartments on a city block are street traffic and possible noise disturbances. In many cases, architects will have to weigh the benefits of orienting the apartment to the sun (despite the exposure to the street and street noise) against the benefits of orienting it to the quiet back courtyard, which may be on the shady side of the building. Here it is also advisable to have floor plans extend to both sides of the building in order to meet all needs.

Parking spaces

Parking spaces are often arranged parallel, diagonal or perpendicular to the street in front of the city block. An attractive green cityscape can be created by breaking up this pattern through trees planted at regular intervals of about five to ten parking spaces. Due to the increased volume of traffic in modern cities, the available parking spaces will probably be insufficient to cover all needs. If this is the case, underground garages located under the buildings or the interior courtyard may be necessary to provide adequate parking. However, underground garages can place restrictions on the design of the open planted areas in the courtyard, and they can substantially increases costs. It is important not to place above-ground parking spaces inside the city block since this not only impairs the visual effect of the courtyard but also causes noise problems and conflicts with the otherwise quiet uses of the rear area.

HISTORICAL EXAMPLES

City blocks in Antiquity

Ever since Antiquity, the city block has been one of the most important elements of urban design. It was used in Greek cities as early as the 6th century BC, and in the 5th century Hippodamus designed the newly founded city of Miletus on the basis of a regular orthogonal grid pattern. › Fig. 14 A large number of Greek colonies, including Olynthus, Agrigento, Paestum and Naples, were also laid out using the block system.

Roman town planning adopted the grid principle and applied it rigorously to its newly founded towns – Cologne, Trier, Nîmes, Bologna and Florence. These settlements usually evolved from a military camp, the *castrum*, and their backbone was formed by two main streets intersecting at right angles, the north-south *cardo maximus* and the east-west *decumanus maximus*. These axes divided the city into four areas › Fig. 15 giving us the term "city quarter". The market and important public buildings were

\\ Tip:
To ensure a high degree of flexibility in the event of mixed uses, the height of a ground floor ceiling on a block can be slightly raised, particularly along main streets. A height of 3.0 or 3.25 m can replace the usual height of about 2.5 m required for residential purposes.

\\ Tip:
As a rule of thumb, if the buildings on a block have a maximum of three storeys, it can be assumed that continuous diagonal parking along the street can meet parking requirements (one parking space per housing unit). However, if the buildings have more storeys, other solutions are required.

Fig.14:
Map of the city of Miletus, Greece

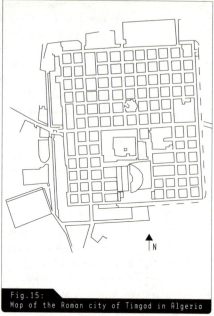

Fig.15:
Map of the Roman city of Timgad in Algeria

located at the intersection point of the main streets, and additional side streets were laid out parallel to them, creating block structures. Deviations from this grid pattern were caused by both special topographic characteristics of the cities (hills, rivers etc.) and the existing streets that were incorporated into the urban network. Difficult topographical locations could thus feature blocks shaped as triangles or different types of rectangles (polygons).

Medieval city blocks

In many places the layouts of Roman cities survived the massive decline in population and urban decay of the post-Roman period, before being revived in the Middle Ages. Although new buildings were erected on the Roman grid patterns, the street layout and block structure remained largely unchanged. Most of the new medieval towns and expanded urban areas that were not based on the Roman grid used a system of polygonal blocks of different shapes and sizes. This created a distinctive public urban space consisting of streets, paths and squares that provided access to the buildings and ensured social living and commerce. Contrasting with this were the private rear areas made up of auxiliary buildings, courtyards and gardens. › Fig. 16

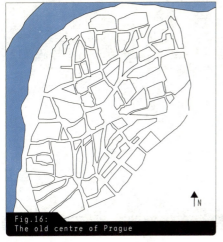
Fig.16:
The old centre of Prague

Fig.17:
Layout of Mannheim (c. 1824)

Cities in colonies

The new cities of the Renaissance (such as the fortress town of Palmanova, founded in 1593 northeast of Venice) and urban planning in the Baroque period (e.g. 17th-century Mannheim) adopted the model of regular grid patterns from Antiquity. > Fig. 17 The same is true of cities founded in North and South America. Spanish and Portuguese conquerors imported the idea of regularly laid out city blocks to the New World as a formal principle of urban design. These basic patterns have survived as central organizational structures to the present day in such cities as Mexico City, Lima, Caracas and Santo Domingo. One famous example in North America is Manhattan. Founded by Dutch immigrants, it uses a chessboard-like pattern as a basis for its urban layout. However, its current cityscape with high-rises and skyscrapers differs radically from traditional city blocks, where buildings are not as tall.

Cities in the industrial age

The rapidly growing cities of the 19th-century industrial age adopted the block structure because of its many advantages: integration into the city as a whole, highly diverse uses, and high level of structural and population density.

During the expansion of Berlin in the German *Gründerzeit* (1871–1914), town planners went so far as to build densely developed blocks with multiple rear courtyards that were accessed from the street through entrance drives. This combination of a very small section of street with a deep plot and high building density facilitated better utilization of the

Fig.18:
Gründerzeit block structures in Prenzlauer Berg, Berlin

available land. However, due to these dense developments, many rooms in the rear courtyard apartments – which housed up to 15 people – did not receive direct sunlight or even sufficient daylight. Because of the population density of far more than 1,000 people per hectare, the hygienic conditions were totally inadequate in most cases, and the living conditions were catastrophic. Tuberculosis and other epidemics were widespread. › Fig. 18

As early as the 19th century, these poor social and hygienic conditions were sharply criticized by many, including Friedrich Engels in his 1845 work *The Condition of the Working Class in England*. In the early 20th century, this criticism led to a partial reform of the city block. In "modern" blocks, planted areas replaced the buildings once erected in the rear courtyards, as illustrated by the projects of Hendrik Petrus Berlage in Amsterdam, J.J.P. Oud in Rotterdam, and Fritz Schumacher in Hamburg.

In the 1920s, the exponents of *Neues Bauen* › Chapter The row fought to have the closed blocks replaced by the freestanding ribbon as a major structural element of the city. › Chapter The ribbon This development fundamentally changed the appearance of European cities. For several decades afterward, the city block became considerably less important as an urban element.

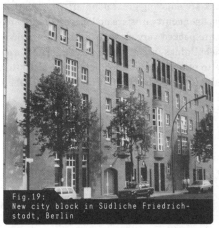 

Fig.19:
New city block in Südliche Friedrichstadt, Berlin

Fig.20:
Quiet communal areas inside the blocks in Südliche Friedrichstadt

The renaissance of the city block

It was not until the 1960s and then the 1970s and 1980s that the city block made a comeback in France, Italy, Germany and other European countries. The catalyst was the criticism many levelled at the destructive effects that modern architecture and *Neues Bauen* had had on cities. The projects built for the International Building Exhibition in Berlin (IBA) between 1980 and 1990 were an expression of this changed philosophy, which also went under the heading of "city repair." › Figs. 19 and 20

Today the city block is once again an important tool in the urban planning repertoire, and even the *Gründerzeit* neighbourhoods that were

\\ Note:
In 1966, the architect and theoretician Aldo Rossi published an influential book entitled *L'architettura della città* (English edition: *The Architecture of the City*). In this work Rossi emphasizes the role block structures play in creating urban space. He also stresses the continuity (permanence) of such structures and their importance for societies, social identity and history.

criticized just a few years back are enjoying great popularity due to their urban density and mixed uses. Rear courtyards are no longer being cleared of all buildings, even if they are packed very close together. Rather, owners are often converting these buildings into attractive and unusual forms of housing and putting them to other uses (studios, lofts, non-disruptive small business). New structures are even being built inside blocks where there is sufficient space.

\\ Note:
The dense block structures of 19th-century industrial cities have gained a new appeal because population density has declined substantially while development density has remained the same. Nowadays it is not uncommon for two people to live in a three-room apartment that a century ago may have housed 25 to 30 individuals.

THE COURTYARD (INVERSE BLOCK)

In terms of urban organization, the courtyard can be seen as the inversion of the city block. The formal arrangement of buildings can be identical for the block and the courtyard, but while buildings on the block (i.e. those along the edge of the block) are accessed from the outside, the buildings of a courtyard are accessed from the inside. Consequently, the front sides of the courtyard buildings are oriented toward the inside space and the rear sides to the outside. The interior area becomes – at least partially – a public space.

As used in urban planning terminology, the term "courtyard" derives from typological models such as the enclosed farmyard or monastery complexes in which the buildings are grouped in cloisters around a courtyard. The term therefore refers to an ensemble of buildings with an open area that is central to its formal and functional organization. As a whole, these architectural ensembles have a self-contained, introverted character.

FORM AND SPATIAL STRUCTURE

Courtyards are usually designed as a complete unit. Their layout is largely based on the principle of neighbourly and collective existence.
> Fig. 21

Outer boundaries

Courtyards can be enclosed by very similar buildings, or they can consist of a group of buildings with different formal designs. In both cases an important feature is that the edges of the courtyard are largely closed off spatially. Aside from entranceways and entrance drives, no large gaps should remain that disrupt the detached quality of the courtyard. If the

\\Note:
The word *close* is also used in English-speaking countries. It originates from *claustrum*, the Latin word for cloisters (monastery), which means "closed off." The German word *Klause* can be used in this context as well (a building or group of buildings closed off from the outside; a hermitage).

Fig. 21:
Different courtyard forms

buildings themselves do not form boundaries, they may be created by other "edge-making" elements, such as walls and hedges.

Like the city block, the courtyard can have entirely different geometric forms. As a forecourt or entrance court, for instance, it can also function as a sub-element in a block structure.

Since the front sides of courtyard buildings face inwards and the exterior sides overlook public space (provided the courtyard is not entirely surrounded by other buildings), the facades on both sides must fulfil specific design requirements. In contrast to the city block, the high degree of formal and creative control required for the courtyard interior does not permit a great deal of freedom for random, unplanned or unauthorized installations and extensions. In a courtyard, there is little distinction between front and back – or outside and inside – in the design of facades and the use of materials.

FORMATION OF URBAN SPACE

Semi-open spaces

The courtyard is detached from the integrated system of public streets and access routes. While it is usually accessible to the public (or else the buildings could not be entered), it constitutes a space with a limited public character that can best be described as semi-public. The design of the transitions between urban space and the courtyard are particularly important and can make use of spatial constrictions, height differences created by ramps and stairs, as well as entrance drives. Architects can also incorporate different ground coverings, planting and other features.

Courtyards are not intertwined as tightly with the urban environment as city blocks, and they are less suited for urban integration. › Chapter The city block The entrances are often pathways that terminate in dead ends and that deliberately do not continue the urban network due to the desire for introversion. The courtyard remains a small world in and of itself. Using a spatial sequence of courtyards that ultimately lead to public street space, this urban element can be better integrated into its environment. The courtyard then becomes a kind of arcade. › Chapter The arcade

FUNCTIONS, ORIENTATION AND ACCESS

Collective use

The courtyard often serves as an urban design model for collective (or cooperative) housing. It offers residents a point of reference and a centre for creating spaces with a degree of privacy and tranquillity, removed from the hustle and bustle of the surrounding city. The courtyard forms a partially autonomous unit within a neighbourhood. This can enhance the residents' sense of security and their ability to monitor the collective space, since the people who live and work in a courtyard will know each other and immediately notice strangers. By orienting important elements toward the outside (such as access routes, open areas and even common spaces), architects can accentuate the courtyard's claim to being a social space. Combined uses are also possible. Special functions and non-disruptive service businesses such as offices and medical practices can be integrated into this urban element.

The courtyard faces similar orientation and lighting problems to the city block. And here, too, many variations have been used in the design of corner areas. › Chapter The city block The inner corner is a special challenge since the rather small front of the building facing the courtyard corresponds to a large exterior area on the garden or rear side.

High degree of development density

The courtyard allows urban land to be optimally exploited for architectural purposes. In combination with the city block, it is often used to enhance building density. Since it is entered from the inside, it can provide access to additional land in the very rear of the plot.

Beyond its access-providing function, the inner space of a courtyard can be a place of shared exchanges, a playground for children, a meeting spot, a storage space for bicycles and prams, a delivery zone, a shared park and recreation area, and much more. If possible, parking spaces for cars should not be located in this inner space to avoid undermining its recreational quality and residential tranquillity. They should be placed outside courtyards or in an underground garage beneath them. > Chapter The city block

HISTORICAL EXAMPLES

Courtyard houses

Houses structured around one or more courtyards were first built in the ancient world and can still be found today, particularly in the Mediterranean region. Such structures were especially widespread in Islamic architecture. However, from an urban planning perspective, this type of courtyard house must be seen as a borderline example of the courtyard as urban building block since it is usually constructed on a single piece of land.

Farmhouses and monasteries

As additional historical examples, we can point to the farmhouse and monastery complexes found in many regions that are for the most part closed off spatially to the outside. What both monasteries and farmhouses have in common is that they are not built solely for residential purpose. The protection they provide from the outside and the shielded social space they create within are important aspects of each. The Certosa in Pavia in northern Italy, an extension of a monastery that provides housing for monks, has a close resemblance to a courtyard structure. > Fig. 22

Charitable housing projects

The residential complexes that Jakob Fugger built for the poor in Augsburg in the period around 1520 can be regarded as an example of communal courtyards from early modern times. They are based on the *hofjes* that were built in the Middle Ages, particularly in Dutch cities. These

\\ Tip:
If an underground car park is required, it can make sense to raise the height of the courtyard by about 1 m in relation to the surrounding area. This will shorten the length of the car park's entrance ramp and permit natural ventilation through the exterior walls.

\\ Note:
The *Wiener Gemeindebauten* in "Red Vienna" was the Social Democratic government's response to the housing shortage among the working population. As part of an extensive construction program launched in 1923, it set out to build up to 30,000 apartments annually. The *Wiener Höfe* (Viennese courtyards) emerged – monumental housing projects with high ceilings, shared courtyards and many subsequent housing facilities. The best known is the Karl-Marx-Hof, which contains more than 1,300 housing units, numerous businesses and communal facilities.

Fig. 22:
Courtyard in Certosa di Pavia

charitable facilities, which date as far back as the 13th century, were often set up as foundations that provided housing for needy groups in society, including elderly people, the poor, the sick and orphans. One of the best-known examples is the Begijnhof in Amsterdam.

Social movements have repeatedly taken up the courtyard concept since it guarantees a minimum amount of shared open space and also offers a degree of privacy despite its high density. Nineteenth-century industrialists used this urban building block as a paternalistic housing model for their workers for the same reason. Even the sprawling "Wiener Gemeindebauten" (Viennese communal housing blocks) built in the 1920s were based on the courtyard; and Michiel Brinkman made use of a similar concept when designing the large city courtyard in Rotterdam-Spangen that accommodated some 270 families (built between 1919 and 1922). Apart from entrances on the ground floor, a wraparound outdoor corridor on the second floor provided access to the apartments. › Fig. 23

Courtyards in the garden city

In the early 20th century, courtyards surrounded by terraced houses were used by the garden city movement as an architectural model for quiet,

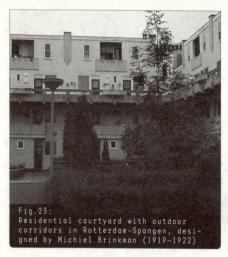
Fig.23:
Residential courtyard with outdoor corridors in Rotterdam-Spangen, designed by Michiel Brinkman (1919–1922)

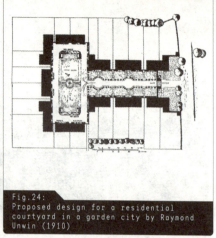
Fig.24:
Proposed design for a residential courtyard in a garden city by Raymond Unwin (1910)

group-based housing on planted grounds. Here, the courtyards were called "closes". Neighbourhood familiarity and small-town identity played a special role. The architect Raymond Unwin designed attractive examples of "closes" in the garden cities of Letchworth, Welwyn Garden City and Hampstead Garden Suburb in the south of England. › Fig. 24

Courtyards continue to be used in the design of communal housing projects, particularly in experimental or cooperative housing construction.

\\ Note:
In his work *Town Planning in Practice*, which was first published in 1910, Raymond Unwin describes the functional characteristics and design features of residential courtyards in new housing estates. He also refers to their economical use of the site and the broad vista that residents in the surrounding buildings have of the planted square and open areas.

THE ARCADE

The arcade has generally evolved as a roofed-over street, lined with shops and businesses, that leads between lines of buildings from one place to the next. The arcade is structurally related to the courtyard in that it is accessed from the inside.

FORM AND SPATIAL STRUCTURE

Glass-covered streets

In most cases, the arcade is a shopping and commercial street covered by a glass roof. As a public path, it is usually accessible only to pedestrians. It is enclosed on both sides by the facades of the adjoining buildings, which are usually carefully designed to create a prestigious impression. Merchandise displayed in the shops on the street is visible to all from behind large windows.

Arcades can be straight, angular or curved. They can take almost any possible linear form or branch off in different directions. The broader spaces at these intersection points can create small squares where people can linger. > Fig. 25

Arcades may run between two different buildings (and sometimes even have several floors). They may also take the form of public paths running through compact block structures. Where this is the case, a great deal of attention is usually paid to the design of the inside facades. In such cases the interior facades are usually carefully designed to reflect the formal characteristics of the exterior.

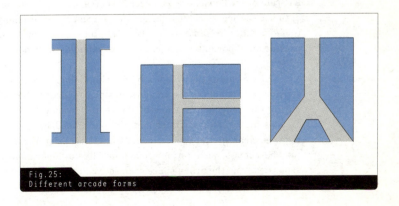

Fig. 25:
Different arcade forms

Fig.26:
Galleria Vittorio Emanuele II in Milan

FORMATION OF URBAN SPACE

Network of paths

The arcade connects paths in the city. This is especially true of the well-known 19th-century arcades in Paris, Brussels, London, Naples and Milan. For instance, the famous Galleria Vittorio Emanuele II in Milan is the shortest route between two important city sites – the cathedral and La Scala opera house. > Fig. 26 Sometimes arcades can create shortcuts that optimize the access network of the city, at least for pedestrians. The glass-roofed arcades in Hamburg, which offer protection from the weather, still provide an important secondary access system to the centre of town.

Climate buffer

As a result of the arcade's roof, a climate buffer emerges inside that considerably enhances the arcade's appeal as a place to tarry, particularly during the inclement seasons of the year. Nowadays a large number of arcades are heated in winter and air-conditioned in summer – but this requires clear spatial separations. Where these exist, the arcade assumes the character of an interior space much like a department store and is separated from the urban space of the city, although this detracts from its image as a continuous covered street space.

Fig. 27:
New arcade in the centre of Aveiro, Portugal

FUNCTIONS, ORIENTATION AND ACCESS

Arcades are primarily influenced by "economic" considerations since in prime inner city locations they can also provide accessibility to the interior plots on a block. Their walkways are largely flat and level so as not to disturb strollers or divert their attention from exhibits and displays in shop windows. › Fig. 27

Commercial uses

Retail and commercial uses predominate and are in some cases supplemented by restaurants. Housing is an exception, but if apartments are included in the arcade, the roof glazing generally begins below the apartments in order to protect against fire, or for lighting and ventilation. Thus it can be said that housing only influences the space of an arcade if the building and apartment entrances are located in the interior.

A key factor, not only for the arcade's commercial success, is functional integration into the urban environment. Both the front and rear of the arcade are important. Architects must make sure that the various entrances lie on lively, busy streets, for if this is not the case, it will cause

Fig. 28:
Hustle and bustle in the Cairo bazaar

Fig. 29:
The 19th-century Kaiserpassage near the Frankfurt railway station

an imbalance between an attractive front area and a less attractive rear area.

HISTORICAL EXAMPLES

Forums and bazaars

Ancient Rome had spatial structures that can be seen as the forerunners of arcades. The Forum Iulium and the Trajan's Market were both lined by shops and businesses, and the street between them was a place for people to linger and conduct business. The Persian city of Isfahan also has a very similar spatial structure. The bazaars and souks in the Islamic world continue to use this organizational principle to the present day. Booths are rowed together on both sides of a central path for the display of various goods. › Fig. 28

Arcades in the 19th century

Arcades became fashionable in European cities such as Paris, Milan and Vienna in the 19th century. They provided the affluent middle classes with a place to stroll that was protected from the weather – and a space that was removed from the noise and dirt of the street. › Fig. 29 The arcades served the economic interests of vendors by making entertainment and pleasure the focus of their design, organization and presentational techniques. Yet they also had a representational function for the wealthy middle classes and the city itself.

Shopping centres and malls

The model of the arcade as a locus of commerce and presentation can be seen as informing the development of present-day shopping centres and large shopping malls. But in contrast to their historical predecessor, these structures are not integrated into the surrounding city. From the outside, they are uninspiring, nondescript "boxes" that usually negate the characteristics necessary to create urban space. › Chapter The shed

\\ Note:
The most important work on arcades is *Passagen-Werk* by Walter Benjamin (English edition: *The Arcades Project*). It consists of literary and architectural reflections on the arcades of the 19th century. The diverse commentary creates a broad impression of the aesthetic design and the economic and functional importance of arcades as social arenas and places of commerce. Benjamin also addresses the physical aspect of walking, describing the arcades as places where strollers observed the displayed goods and services partially out of "scientific" interest and partially for amusement.

\\ Note:
In the book *Project on the City: Harvard Design School Guide to Shopping*, published in 2001, a group of writers under the architect and theoretician Rem Koolhaas associatively examine the historical development of shopping streets and centres, juxtaposing a series of images of Roman fora and Persian and Arab bazaars with today's shopping centres and malls.

THE RIBBON

Ribbons (*Zeilen*) are linear, freestanding urban elements that are deliberately oriented away from the street space to achieve "hygienic" advantages such as the best possible exposure to light and ventilation. They were developed in the 1920s as a reaction to the overcrowded urban space created by the block structures and corridor streets of the traditional city. They can thus be understood as a critique of living conditions in the tenements constructed in the late 19th century. › Chapter The city block

FORM AND SPATIAL STRUCTURE

The ribbon can be seen as a further development of the row. However, in contrast, it is not designed to form a bordered street space. In most cases only its "head," or short side, is oriented toward the access street. This independence from the street vector allows the ribbon to be oriented to achieve maximum exposure to sunlight.

Ribbons are thus not parallel but perpendicular to the street and are accessed via a secondary footpath (in some cases a cul-de-sac). › Fig. 30 Access to ribbon developments is usually from the side less favoured by the sun, i.e. the east or north side (in the southern hemisphere, the south

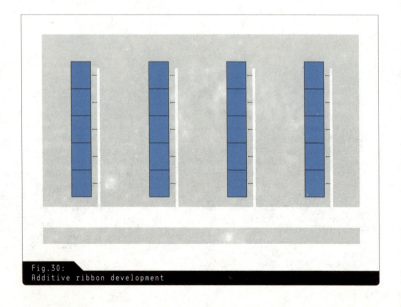

Fig.30:
Additive ribbon development

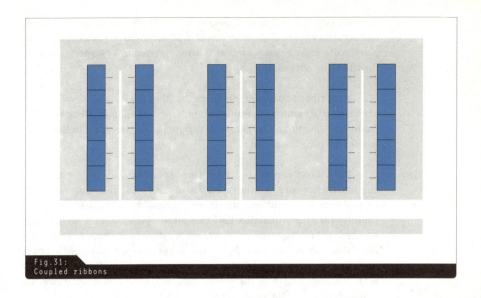

Fig. 31:
Coupled ribbons

side). This prevents access paths from disturbing the sunnier southern and western sides. These sides are commonly used for private open spaces such as balconies, loggias, roof terraces and, on the ground floors, small garden plots.

Additive ribbons

The additive repetition of this pattern in ribbon developments creates an extensive urban structure in which the front side of each ribbon faces the back of the adjacent one. This direct juxtaposition of rear (private) and front (public) spaces can result in a lack of spatial clarity, although this conflict can be mitigated by delimiting ribbons from one another using vegetation, varying structural levels and outbuildings such as bicycle and storage sheds.

Coupled ribbons

Another possibility consists in a mirrored rather than an additive arrangement of ribbons and their access paths. The result is a series of "coupled" ribbons with the open spaces between the ribbons containing either adjacent front areas or adjacent rear areas in an alternating pattern. › Fig. 31 This arrangement means that apartments and in particular their adjoining open spaces have different orientations to the sun, but it has the advantage of lending the external space a social character.

Ribbons can be composed of single-family houses joined in a line (in the form of terraced houses of between two and three storeys) or joined multiple-family dwellings (connected apartment blocks of between three and six storeys). The slab structure used for large unitary residential complexes with eight or more storeys is often referred to as a special form of the ribbon development.

Most ribbon structures are linear, but they can also be curved, angled or consist of a number of sections set off from one another. Along with differences in length and height, such variations can provide a rudimentary tool for shaping urban space.

Standardization

The ribbon can be understood as typical of the age of mass production. Its linearity and the repetition of individual units make it highly amenable to the use of industrially prefabricated elements. However, the capacity for standardization that this allows (which makes for a highly economical building process) can mean a risk – in the case of persistent repetition – of monotonous forms and urban designs. An example is the industrially prefabricated high-rise apartment blocks built on many large-scale housing estates in the second half of the 20th century, which are found above all in Eastern Europe.

FORMATION OF URBAN SPACE

The orientation of ribbons toward light and the sun results in their almost complete independence from the surrounding urban space and the local network of access streets. In this sense the ribbon negates traditional concepts of urban form and space. This independence often results in it becoming an anti-urban element that makes no claim to a spatially formative role in the conventional sense. This is particularly problematic when ribbon developments – which originated as structural components of city peripheries – are built on inner-city wasteland and in the gaps between buildings.

Flowing space

In cases where the areas between ribbons are not enclosed, the result is usually a flowing surrounding space that lacks clearly defined public and private areas. Usually covered by a lawn or other vegetation, this homogeneous space is in principle supposed to function as a communal area, particularly where the ribbon structure is made up of apartment blocks. However, such communal areas are rarely utilized in practice. Instead they remain anonymous spaces for which no one feels responsible and they quickly fall into neglect. › Fig. 32 Furthermore, due to the lack of defined street space, these open areas are subject to only a low level of social control, which can contribute to residents' feelings of insecurity, particularly in the case of large apartment blocks.

Fig.32:
Neglected spaces between the ribbon structures of a modern housing estate

Fig.33:
Low transverse buildings that can be used for commercial purposes close off the space between ribbons from the street.

FUNCTIONS, ORIENTATION AND ACCESS

Residential use

As an urban building block, the ribbon accords with the concepts of functionalist urban design, which – as seen in the Athens Charter (1933) – strictly separates housing, work, transport and recreational functions. › Chapter The row For this reason ribbons are generally only used for residential buildings, with office and commercial uses being the exception. Due to its deviation from the access street and thus the lack of a direct view of the building from passing traffic, the ribbon does not readily lend itself to such public uses.

This is also the reason that small shops and other providers of daily services are sometimes located in the short front section of the ribbon, directly on the access street or in low intermediary buildings between the ribbons. › Fig. 33 These structures have a dual urban-planning function. They restore a degree of continuity to the public street space, and they also shield the areas between the ribbons both spatially and acoustically from the street. As a result the urban space on both sides becomes more clearly defined – a first step back toward shaping urban space. A mixed form is produced that combines the ribbon with the (open) block and includes quiet, semi-public spaces.

Orientation

As in rows and blocks, apartments in ribbon developments can be differentiated into those with an east-west orientation and those with a north-south orientation. An east-west orientation has the advantage that recreational rooms receive sunlight from both sides, whereas a north-south orientation provides natural light from only one side. For this reason, orientation is an important factor when considering building depth and how open floor plans should be. › Chapters The row, The city block

Residential pathways

Since ribbons are connected to the street network only on their short sides, building entrances are usually accessed via footpaths along one side. However, in some cases we find an alternating pattern of roadways and footpaths such that a ribbon can be accessed on one side via a cul-de-sac and on the other via a footpath. This in turn has an effect on the internal organization of ribbon developments in terms of the location of main entrances and individual rooms within the apartments and the orientation of open areas. The advantage of this pattern is that it allows for a separation between motorized and non-motorized traffic. The footpaths leading to the ribbons often take on a semi-public character, which, outside entrance doors, promotes recreation and communication, say, between playing children.

Outside areas

For a long time ground-floor residents – whose apartments tend to be slightly elevated above ground level – were not permitted to use any outside areas directly in front of their apartments. Appropriating the green areas in front of apartments was regarded as contrary to the principle of "equal rights for all". Only recently has it been recognized that residents do not necessarily share the same interests in this regard. Some would like a small garden while others prefer a balcony or roof terrace. Furthermore, allowing ground-floor residents to have a small garden or terrace area leads to an improvement in the aesthetic quality of outside spaces, a greater feeling of responsibility for their care, and better social supervision. This benefits the general security of the whole residential community.

HISTORICAL EXAMPLES

The ribbon is a comparatively new urban building block. With the exception of a few historical forerunners, such as the Adelphi development in London, which was built between 1768 and 1772 by the brothers James und Robert Adam, or the blocks with outdoor-corridor access that were built in northern Italy in the late 19th century, the ribbon development was essentially a creation of the *Neues Bauen* movement of the 1920s.

Residential estates in the 1920s

The most prominent example is the Dammerstock residential estate, which was built in 1927–1928 in Karlsruhe as an exhibition project. › Fig. 34 The final version of the site plan was designed by Otto Haesler and Walter Gropius, who was director of the Bauhaus at the time. The plan was a deliberately provocative manifesto for a completely new type of urban structure. It proposed a strict north-south ribbon vector with an east-

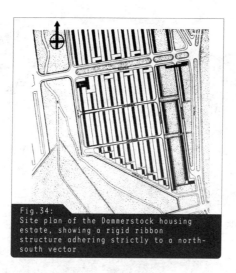

Fig.34:
Site plan of the Dammerstock housing estate, showing a rigid ribbon structure adhering strictly to a north-south vector.

west orientation for all apartments, a seemingly endless linearity, identical intervals between the individual ribbons and the abandonment of all conventional notions of spatial organization. There is probably no other urban-planning project that has created such a degree of controversy from its inception. While some saw the estate as an embodiment of modern, progressive urban development, providing optimal living conditions (light, air, sun) for everyone, others denounced its stubborn adherence to the principles that informed the abstract estate ground plan, the uniform architecture and lack of spatial formation. › Fig. 35

\\Note:
Blocks with outdoor-corridor access are elongated ribbon developments containing connected residential units. They are accessed on one or more levels by a shared outside walkway.

\\Note:
The representatives of modern architecture focused on the ribbon primarily because of the progress it made in urban "hygiene" and because of the social implications it had for life in an egalitarian society that offered the same residential and living conditions for all. They also emphasized the economic advantages this urban element offered for the mass production of building elements. The theory of ribbon development was systematically and comprehensively discussed and documented for the first time within the context of the "Rational Site Planning" segment of the CIAM congress held in Brussels in 1930.

Fig.35:
Stringently organized ribbon development in the Dammerstock housing estate in Karlsruhe

The model established by the Dammerstock residential estate inspired numerous other well-known *Neues Bauen* projects in the late 1920s, such as the Hellerhof estate (1929–1932) and the Westhausen estate (1929–1931) in Frankfurt, Siemensstadt (1929–1932) and Haselhorst (1928–1931) in Berlin, and the Rothenburg estate in Kassel (1929–1931). In the 1950s and 1960s, this urban-planning concept was taken up across Europe and in other parts of the world as the blueprint for a standardized form of residential building for lower-income earners.

Modernization programmes

In the 1970s, the ribbon development as an urban component became the target of the postmodern critique of functionalist architecture in general. It fell into disrepute primarily because of the social problems engendered by the predominance of economically disadvantaged residents, its functional inadequacy as a so-called dormitory town and its aesthetic monotony. However, since the 1990s, modernization programmes and design improvements (the addition of generous balcony spaces, demolition of buildings that are too high, remodelling of the surrounding environment) have succeeded in improving residential conditions in many ribbon development neighbourhoods.

THE SOLITAIRE

In urban-planning terms, a solitaire refers to a building that either stands alone or is clearly distinguishable from its urban surroundings. Freestanding buildings such as granges, farmhouses, castles and monasteries have been part of the cultural landscape since time immemorial. However, in the more densely built context of the city, solitaires initially tended to be defined by the fact that they protruded from the regular network of urban elements constituted by rows and blocks. They were usually public buildings (temples, churches, town halls) or the buildings of the ruling classes (castles, fortresses). They were later built as residences for the rich (villas, palaces) or to house the growing urban infrastructure (schools, theatres, opera houses, museums, hospitals, parliament buildings, universities etc.). Today, the solitaires found in large cities include residential and office towers as well as freestanding, single-family homes that use up an increasing amount of land.

FORM AND SPATIAL STRUCTURE

Solitaires are quite distinctive from the surrounding buildings in terms of their size, importance, geometry, architectural design and construction materials. In cases where they are not spatially separated from neighbouring structures, their distinctiveness in terms of form and decoration make them clearly recognizable as self-defined structural units.
› Fig. 36

Formal autonomy

Where a solitaire is not connected with any other buildings, its design can be relatively independent from the urban-planning context in terms of form and proportions. This means that architects have far more creative latitude than when designing other urban building blocks, with the result that solitaires can take the form of slabs, towers, cubes, cylinders, pyramids, and a range of hybrid combinations. Nevertheless, if they need to be integrated into a larger urban ensemble or have a specific relevance for the urban silhouette or the landscape, the size, form and facades of solitaires should accord with certain design principles.

FORMATION OF URBAN SPACE

The design concept for a solitaire does not seek to establish any direct connection with the buildings around it. In many cases, the aim is to create a structure that is obviously distinct from its urban framework and creates a particular focus in the cityscape and a specific spatial effect.

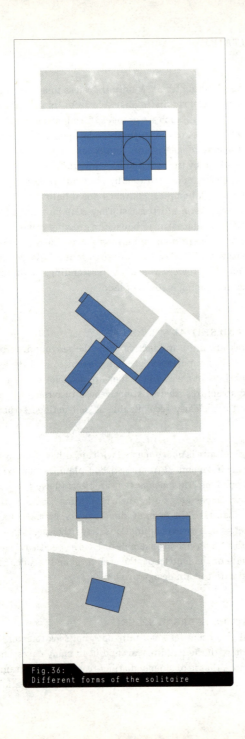

Fig.36:
Different forms of the solitaire

Fig. 37:
The Temple Mount in Dougga, a town in Tunisia occupied by the ancient Romans

Exposed location

In some cases, solitaires are deliberately disconnected from the urban framework and placed in a prominent location. Examples of this can be seen in the temples and shrines of Antiquity. Such buildings do not specifically shape the urban space. Rather, they form vivid, sculptural culminations of an overall urban context, which they accentuate and orchestrate. Placing the structure in a topographically prominent location can strengthen this effect, as illustrated by the Acropolis in Athens and many other church and religious buildings. › Fig. 37 In the Baroque and absolutist periods, and later in 19th-century cities, town planners gave solitaires particular emphasis by building them at the ends or intersections of important thoroughfares and visual axes. › Fig. 38

Integrated location

However, the lack of space in densely built cities often means that solitaires do not stand completely alone. They are often spatially integrated into the side of a city square, a building line or a building group. › Fig. 39 This is particularly evident in dense, compact medieval cities, where large cathedrals, town halls, convents and even tithe barns are integrated into

Fig. 38:
The Madeleine in Paris is located at the intersection of the street and visual axes.

Fig. 39:
Santi Giovanni e Paolo as part of the ground plan of Venice

the urban framework while remaining distinct from their surroundings due to their size, facade design and particular position within the city layout.

Spatial effect — In the cities of modernity, solitaires were usually built as completely freestanding buildings whose orientation allowed optimal lighting and ventilation. › Chapter The ribbon This was also a result of changed concepts of urban space, which, according to the objectives of the *Neues Bauen* movement, needed to be open and flowing rather than closed in the traditional manner. In this type of an urban space, which is characterized above all by the interplay between freely placed individual buildings, solitaires can have a pronounced spatial-sculptural effect. › Fig. 40

FUNCTIONS, ORIENTATION AND ACCESS

Functional specialization — In principle the solitaire can take on a range of functions as an urban building block. Although in the case of large buildings mixed uses are possible (e.g. commercial uses on the ground floors of residential high rises), the solitaire is usually characterized by a high degree of functional specialization, which encompasses specific public functions (town hall, community centre, school, university, museum etc.) and private uses that the design seeks to accommodate through a specific architectural identity (residential complex, government authority, business headquarters, hotel).

Fig.40:
Le Corbusier's Unité d'habitation in Marseilles: a solitaire serving as a prototype of the "vertical city"

Orientation
    Orientation and natural lighting generally do not present problems for solitaires. If they are not unusually deep or wide buildings, which can result in dark areas in the interior, solitaires receive sunlight and can be ventilated from all sides. Problems with shadows can occur where high-rise blocks are built too closely together. This can be seen in some inner-city locations such as New York and in the mega-cities of East and Southeast Asia, including Beijing, Shanghai, Hong Kong and Seoul.

Car parks
    The open areas usually found in front of, beside or behind the building generally make it possible to locate car parks at ground level. However,

\\ Note:
Le Corbusier's "machines for living" – particularly the Unité d'habitation, which was constructed in Marseilles between 1945 and 1952 for 1,300 residents – are examples of the attempt to create "vertical cities" that provide living space in green surroundings for a broad range of income groups. In his essay collection *La Ville Radieuse*, published in 1935, Le Corbusier describes multi-storey buildings that appear to float on pylons above the ground, and that make nature an integral element of the living space. These solitaire residential buildings were envisaged as completely independent of their urban surroundings. As a consequence, their internal structure is complex and includes not only apartments but also shopping streets, communal spaces, a hotel, kindergartens, roof terraces and sporting facilities.

this may conflict with other proposed uses for the surrounding space (recreation, social interaction). For this reason, underground car parks are preferable, particularly in high-density residential and public buildings, which may require a large amount of parking at particular times.

HISTORICAL EXAMPLES

Reference has already been made to the individual, freestanding house as a basic building block of settlement structures, primarily in rural and village contexts, as well as to the solitaires associated with religious or secular authorities in ancient and medieval cities.

Palaces and villas

From the 15th century onwards, the palaces of important city residents took on increasing significance as solitaires within the urban landscape. A great deal of money was spent on structures that adequately represented the power of influential families, as can be seen in the Palazzo Pitti and the Palazzo Strozzi in Florence and the Fuggerhaus in Augsburg. However, in the age of absolutism, the significance attached to these patrician buildings shifted to the palaces of the nobility and the seats of royalty. Their sheer size clearly distinguished them from the houses of the urban bourgeoisie with their relatively small structural elements and often Gothic character. In the 16th century, Andrea Palladio created a form of villa architecture in Italy's Veneto region that became internationally renowned for its proportionality, formal expression and charm. It continues to serve as a paradigm today. > Fig. 41

Urban infrastructures

The expanding industrial city of the 19th century produced an array of new functional requirements as regards commercial, cultural, social, political and transport infrastructure. The prestigious solitaires – market halls, department stores, theatres, opera houses, museums, educational institutions, hospitals, parliament buildings, railway stations and many

\\ Note:
In his famous *Four Books on Architecture* (1570), a standard work on architectural theory, Palladio describes his villa buildings as examples of rural architecture, yet he does not position them as antithetical to the city.

Fig.41: Andrea Palladio's Villa Rotonda, Vicenza

other types of public buildings – now assumed a high level of urban-planning significance and became a dominant element in urban space.

Single-family homes

The car as individual transport in 20th-century cities contributed to a further quantitative and qualitative leap in urban development. The new level of mobility the car provided led to the extensive construction of solitaires in the form of freestanding single-family homes on the edges of cities and towns, above all in the prosperous countries of the industrialized world. They in turn resulted in the progressive depletion of open space and the destruction of rural landscapes. › Fig. 42 Furthermore, this development generated significant costs for infrastructure such as roads and sewage systems, and meant that a significant proportion of the population had to travel a long way to reach social, cultural and commercial facilities.

Residential apartment blocks

By contrast, in the field of high-rise construction, the solitaire allows for greater density. Even so, at least in the case of apartment complexes, the structural density is not greater than that found in rows and blocks of four to six levels because of the space that must be left between buildings. Furthermore, many residential solitaires tend to lack an urban-planning context, meaning they do not create urban spaces that correspond to the human need for orientation and security. › Fig. 43 When designing high-rise residential buildings, it is therefore necessary not only to select an appropriate location but also to define the intended target groups precisely. While this residential form is unsuitable for families with children, older people and socially disadvantaged groups, it can provide an attractive and

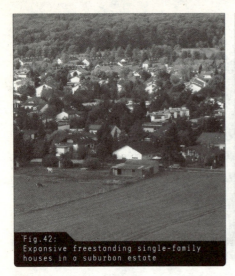

Fig.42:
Expansive freestanding single-family houses in a suburban estate

Fig.43:
Vertical density in a freestanding high-rise complex

exclusive alternative for young professionals, couples, single people and more affluent sections of the population.

City villas

The city villas constructed in recent decades are solitary residential buildings between four and six storeys tall. In urban-planning terms, they are an intermediate form between the single-family house and the high-rise residential building.

Skyscrapers

The skyscraper is a special form of the solitaire. Examples such as bank towers and company headquarters are deliberately designed as spectacular buildings that project a powerful corporate image. However, their effect from a distance is quite different from the one they have close-up. From a distance, skyscrapers can be fascinating, needle-like structures reaching to the sky. They can even form groups, › Chapter The group with a decisive influence on the silhouette of the city, as seen by the skylines of Frankfurt › Fig. 44 and – even more impressively – New York. However, from the perspective of the pedestrian or motorist the same buildings are experienced less as solitaires than as objects that define the street space. For this reason the design of their facades and the relationship between the interior and exterior of the lower levels is extremely important. If possible, these levels should have functions that allow public access in order to inject life into the street spaces of otherwise mono-functional office districts.

Fig. 44:
The Frankfurt skyline

"Trans-
locational"
solitaires

In recent times, the comprehensive mediatization and globalization of our societies has produced a building type that could be described as a "translocational" solitaire. One example is the Frank Gehry's Guggenheim Museum in Bilbao, with which many people are familiar although they themselves have never visited it. › Fig. 45 This museum has become so engrained in the general consciousness that it constitutes a kind of virtual architecture that exerts an influence without being physically present. An analogous effect was no doubt created in earlier periods by structures such

\\ Note:
The 1960s saw renewed discussion of the relevance of striking solitaires for urban design. In his book *The Image of the City*, Kevin Lynch points to the great significance of memorable buildings — which he refers to as markers — for our perception of, and orientation within, urban spaces and structures. In this context, he speaks of the mental maps of familiar urban spaces and buildings that we draw in our mind's eye. These maps are not scale plans but records of our individual experience of urban space.

Fig.45:
The Guggenheim Museum in Bilbao, designed by Frank Gehry

as the Roman Colosseum, the Leaning Tower of Pisa and the Eiffel Tower. However, the significance of such buildings has increased enormously in recent times due to their presence in the media (on television, in advertising etc.) While all of them are remarkable for their size, they also share a specific expressiveness. The construction of the Guggenheim Museum in Bilbao – to return to the first example – has massively increased the number of tourists visiting the city.

THE GROUP

A group is an arrangement of buildings whose character is based more on inner compositional logic than external urban organization. Highly dense and organizationally complex groups are also referred to as clusters.

FORM AND SPATIAL STRUCTURE

Within a group, each element is attuned to the others and can only be understood in terms of its relationship to these other elements. › Fig. 46 As a rule, groups are based on an organizational principle according to which a whole is structured on the basis of interdependent parts. These are not combined additively as in the row › Chapter The row and are thus not arbitrarily extendable.

Manifold spatial configurations

The typological composition of groups can be very uniform, that is, limited to only a few types. However, the group can also combine a very diverse range of building types. It can include the different urban building blocks discussed in this publication (solitaires, ribbons, rows, courtyards and block fragments), which are arranged to create a formal and spatial tension. In these spatial configurations, concepts such as closeness, distance, integration and space (filled or empty) play an important role. Groups can include both open and closed building forms, and they

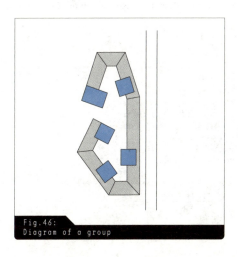

Fig. 46:
Diagram of a group

are often organized around a common centre, an open space, a square, a green area or a spatial sequence of these elements. These spaces take on a particular significance for the identity of the group.

FORMATION OF URBAN SPACE

Dissociation from the urban context

The shared identity and recognizability of the group result in a more or less distinct dissociation from the surrounding urban context. In addition, groups form their own inner urban spaces and spatial sequences that assume different degrees of distinctiveness depending on the size and extensiveness of the group.

On the one hand, the individuality and recognizability of the group can allow inhabitants to identify with their residential and living environment. On the other, there is a danger that groups can become separate "islands" for different population groups, requirements, financial possibilities etc., which can undermine the socio-spatial coherence of the city. Such dangers and their consequences are clearly illustrated in the increasing prevalence of gated communities in large cities throughout the world.

>

FUNCTIONS, ORIENTATION AND ACCESS

In many cases, groups consist of residential building projects, but they can have other functions, too. Examples include universities (where a self-contained campus constitutes a "city within the city"), hospitals and business parks. Mixed uses are possible but tend to be the exception because the group does not often allow adequate integration into the

\\Note:
Gated communities are residential areas that are closed to general access and are guarded. Access is strictly regulated to protect the inhabitants from the dangers of the urban environment such as street crime and burglaries. Naturally, such exclusive residential areas create social segregation, but this is what their inhabitants desire. The concept of gated community can also be used in a figurative sense to refer to social and economic groups that seek to shield themselves from their environment.

surrounding urban environment. For this reason, a group needs to have a certain size if mixed uses are to be sustainable.

One advantage of the group – particularly the more complex cluster – lies in its structural density. However, this can produce orientation and lighting problems and, where residential units are too close together, a lack of privacy for the inhabitants.

Due to the importance of intermediate and internal spaces for the identity of the group, the use of cars in these areas is usually not permitted, or only to a limited extent. This creates attractive recreational and communal areas in the centre of the group, which are restricted to pedestrian and bicycle traffic. Parking spaces are located at ground level in the areas on the edge of the complex, or multi-storey car parks or underground garages can constructed, say, beneath the communal areas.

HISTORICAL EXAMPLES

From an urban-planning perspective, the Minoan Palace built in the first half of the second millennium BC on the island of Crete, and in particular the Palace of Knossos, can both be described as groups. ⟩ Fig. 47 Their complex spatial sequences and high-density created cluster-like structures whose interior orientation and labyrinthine organization ultimately provided the basis for the myth of Ariadne's thread. The interlocking residential quarters of cities in the Arab-Islamic world can also be described as groups or dense clusters.

However, the group is predominantly a product of the recent history of urban development, and social and communal considerations have played an important role in its development. We thus find groups in the workers' housing estates constructed at the end of the 19th century and in the designs of the garden city movement. Today, such projects are commonly seen as a viable means of saving costs and using space economically, and are often associated with an environmentally friendly and community-based approach to building. Groups and clusters are often collective building projects undertaken by construction collectives and building cooperatives. In this context, the construction of groups is not only seen as a way of reducing costs but also of providing a structural and urban expression of communal existence. In addition, there are of course projects financed by private investors and sold as condominiums or single-family houses.

Examples of groups and clusters of a particularly high architectural quality can be found in the work of the Swiss architectural firm Atelier 5.

Fig.47:
Layout of the Palace of Knossos on Crete

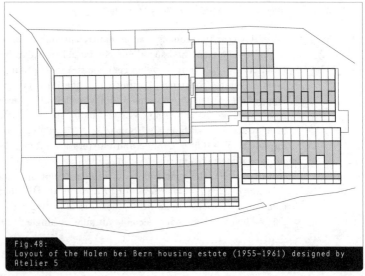

Fig.48:
Layout of the Halen bei Bern housing estate (1955-1961) designed by Atelier 5

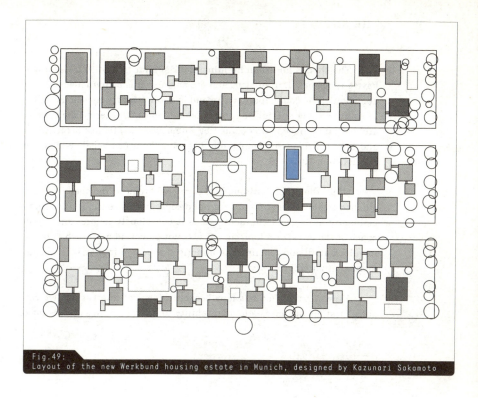

Fig. 49:
Layout of the new Werkbund housing estate in Munich, designed by Kazunari Sakamoto

For years the firm has been developing housing projects as self-contained residential units that project a distinctive identity. A prime example is the Halen bei Bern estate built between 1955 and 1961. › Fig. 48 A more recent example can be seen in Japanese architect Kazunari Sakamoto's 2006 design for the Werkbund housing estate in Munich. The design is made up of a dense patchwork of solitaires of differing heights and a differentiated network of public, semi-public and private open areas. › Fig. 49

THE SHED

The shed is an urban building block similar to a solitaire, and may have a range of different sizes and dimensions. It is a characteristic phenomenon of the contemporary city and is notable for its conscious failure to establish any spatial or contextual reference. The term "shed" as an architectural concept was coined by Robert Venturi, Denise Scott Brown and Steven Izenour in their study Learning from Las Vegas, which was published in 1972.

Of all the urban building blocks referred to here, the shed is distinguished by its abnegation of external design. In this sense, it exhibits an eminently anti-urban character, since it consciously ignores the public space of the city. For a long time it was not perceived as an urban building block at all and remained an unnoticed aspect of industrial and commercial architecture.

However, the shed has now become a focus of interest for two reasons. First, its openness and adaptability to a diverse range of uses make

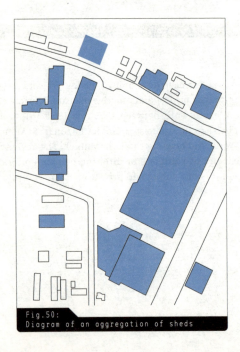

Fig.50:
Diagram of an aggregation of sheds

Fig. 51:
Sheds in the contemporary urban landscape

it economically and structurally attractive. Second, its (non-)design has influenced the appearance of extensive areas of cities and their surroundings and thus the everyday living space of large numbers of people.

FORM AND SPATIAL STRUCTURE

In principle, sheds can assume any form that can be realized structurally, technically and economically. Its geometry and dimensions are also flexible, and shed structures can range from small factories to spacious shopping malls. > Fig. 50 However, the most distinctive characteristic of the shed is its lack of exterior design. The result is that it turns away from its surroundings. Its spatial disposition is derived exclusively from both technical requirements and interior organization and design, which, in contrast to the exterior, often place great importance on an attractive and customer-friendly environment. > Fig. 51

FORMATION OF URBAN SPACE

Sheds can be located anywhere. However, they significantly disrupt urban space because they fundamentally negate the architectural and urban design of streets and public spaces. As a rule, therefore, sheds are found on the outer peripheries of cities and the areas directly beyond them. Nevertheless, they have a significant influence on the everyday lives of a city's inhabitants.

The lack of exterior design is partly compensated for by billboards and large advertising areas, which are used as means to draw attention to the interior life of the shed and to attract customers. > Fig. 52 In some cases,

Fig. 52:
Advertising pillars and billboards used to draw attention to the interior functions of sheds

a particular corporate identity is developed by covering facade areas with recognizable patterns and fragmentary architectural elements.

FUNCTIONS, ORIENTATION AND ACCESS

Functional openness

Sheds can assume practically any function. Their form and spatial structure result from this functional openness and permanent convertibility. Sheds are usually surrounded by sufficient open space to accommodate large parking areas. In the rare case of higher visitor numbers and restricted space, for example in the inner city, additional sheds are erected as multi-storey car parks, or basement garages are constructed. › Fig. 53

Staging the city

In its interior, the appearance of the shed is often dramatically different. In the case of shopping centres, shed interiors are often designed to create the ambience and flair of an inner-city location. A prime example is Main-Taunus-Zentrum, which was constructed in the 1960s near Frankfurt as Europe's first shopping mall, and which can seen as representative of many such complexes. Inviting visitors to stroll and window-shop, it is organized around an inner arcade lined on both sides by shop windows and attractive displays. › Fig. 54 In order to enhance the "city feeling", widened areas resembling city squares have been integrated into the space, along with fountains and sculptures with an antiquated look. Shoppers can also

Fig.53:
Sheds as containers for parking facilities

Fig.54:
Shed interior organized as a shopping boulevard lined with attractive shop window displays

relax in restaurants and ice cream parlours. Thus, the lack of exterior design and the mall's dissociation from the surrounding city is not reflected at all in the interior of the structure.

HISTORICAL EXAMPLES

Sheds in the sense used here first appeared in significant numbers in the period after World War II. Until this point, considerable importance was attached to the architectural design and structural integration of industrial facilities, transport infrastructure buildings and department stores. Notable examples include the buildings designed by Peter Behrens for AEG in Berlin prior to the World War I, Auguste Perret's Garage Rue Ponthieu (1905) in Paris, and the Tietz department store built by Bernhard Sehring on Berlin's Leipziger Strasse.

Abnegation of external design

Economic advantages – and to a certain extent, the level of banality that crept into functionalist building – ultimately led to the practice of giving attention to the design of exterior surfaces only in those cases where utility buildings were exposed to customer traffic. Entrances and interior public areas were designed to have a representational function, while exterior surfaces, having no representational role, were neglected. Today the design of factories, multi-storey car parks, retail outlets and shopping malls often follows this logic.

Las Vegas

The best example of sheds like these can be found on the Strip in Las Vegas, where entertainment facilities, casinos, amusement arcades and hotels are based on the shed principle. Illuminated signs and billboards draw attention to the buildings and become surrogates for architectural and facade design, which are organized solely on the basis of their interior requirements. In their study of Las Vegas, Robert Venturi, Denise Scott-Brown and Steven Izenour refer to these structures as decorated sheds.

This development can be observed throughout the world. Ever greater areas of urban territory, in particular the peripheral areas around them, are changing drastically in appearance due to the influence of sheds. Since many everyday activities, particularly those that have to do with shopping and consumption, now take people into these areas, many spheres of life are being affected by this transformation, which is directed against the city and its public spaces.

Virtual worlds

These changes are being reinforced by telematic developments and the increasing significance of virtual worlds. The architect Bernard Tschumi has argued that the Internet will decisively alter the design and appearance of our cities, citing as proof that he has not entered a bank since the advent of Internet banking. Shopping and interaction with administrative authorities are similarly being affected. Historical, representational architecture which once signalled from the outside how the building was used has become obsolete. A bank no longer has to look like a bank. If no one enters it anymore, it will probably suffice to place it in a shed.

IN CONCLUSION:
FROM BUILDING BLOCK OF THE CITY TO URBAN STRUCTURE

"Building cities involves designing groups and spaces with three-dimensional materials."

Albert Brinckmann (1908)

Once the characteristics and structures of individual urban building blocks have been identified, this knowledge can be applied to larger (urban) spatial contexts. It is only at the level of an urban neighbourhood or an entire city that the networks and interplay between the building blocks actually form the spaces in which we move, live and work every day. Although this book has presented these urban elements individually, it is important to study the diverse connections between them in order to understand the effect they have on the realities of urban existence. It is only in this way that the city can be comprehended as a complex system of spatial, functional and social interdependencies.

Urban building blocks play a central role in this system. As built structures they determine they way individual buildings are used, while indirectly influencing the intermediate spaces of the city – streets and access paths, squares and parks – in which public (and private) life takes place. The resulting spatial integration and forms of functional appropriation by residents and visitors also have an effect on the composition of the urban elements themselves.

Nevertheless, although a thorough knowledge of individual urban elements is essential, the city should be studied and designed as a whole. It is particularly important to bear this in mind when tackling the diverse challenges posed by the technological, demographic, socio-cultural and economic changes currently taking place in cities. Parts of cities are all too often planned and developed as functionally and socially isolated urban fragments. The "islands" that are thereby created may be optimal for certain uses and lifestyles, but other areas of the city may be left behind. In cases where the interplay of urban elements and the overall urban integration do not function, spatial and functional deficiencies often develop that rapidly start to exert an effect on economic and social spheres. The integration of the different parts of the city into an overall urban structure must therefore be a central concern of urban development.

As the 21st century begins, the sheer size of the agglomerations now forming extensive urbanized regions, international metropolises and megalopolises and the associated differentiation of their societies are naturally challenging the validity of long-established concepts and models. Nevertheless, many of the tasks involved in urban development have remained the same. Urban development has to create a physical identity and distinctive functional and social living spaces. It must include the design of intermediate spaces and, in particular, public spaces that are accessible to everyone at all times. It must search for a balance between public and private interests.

In this context, studying the building blocks of the city is a first step towards understanding the built urban structure in terms of its central significance as a physical living and cultural space and in an effort to develop it as such. It is on this basis that the frameworks and ultimately also the methods of city design need to be developed. Engaging in this process enables students to derive practical tips and insights that can help guide them in the task of conceptually planning urban living space. Further study and, of course, professional experience will lead them beyond these primary building blocks of the city to more complex arrangements of urban space. It is in such living and constantly changing contexts that the concepts introduced here will need to prove their worth.

APPENDIX

LITERATURE

Leonardo Benevolo: *The History of the City*, MIT Press, Cambridge (Mass.) 1980

Walter Benjamin: *The Arcades Project*, Belknap Press, Cambridge (Mass.) 1999

Chuihua Judy Chung, Jeffrey Inaba, Rem Koolhaas, Sze Tsung Leong: *Project on the City 2: Harvard Design School Guide to Shopping*, Harvard Graduate School of Design 2002

Friedrich Engels: *The Condition of the Working Class in England in 1844*, J. W. Lovell Company, New York 1887

Robert Fishman: *Bourgeois Utopias. The Rise and Fall of Suburbia*, Basic Books, New York 1987

Ebenezer Howard: *Tomorrow: A Peaceful Path to Real Reform*, London 1898

Le Corbusier: *The Athens Charter*, Grossman Publishers, New York 1973

Le Corbusier: *The Radiant City*, Orion Press, New York 1967

Le Corbusier: *Towards a new Architecture*, Dover Publications, New York 1986

Kevin Lynch: *The Image of the City*, Cambridge Technology Press, Cambridge (Mass.) 1960

Franz Oswald, Peter Baccini: *Netzstadt. Designing the Urban*, Birkhäuser, Basel 2003

Andrea Palladio: *Four Books on Architecture*, MIT Press, Cambridge (Mass.) 1997

Philippe Panerai, Jean Castex, Jean-Charles Depaule: *Urban Forms*, Architectural Press, Boston 2004

Aldo Rossi: *The Architecture of the City*, MIT Press, Cambridge (Mass.) 1982

Camillo Sitte: *The Art of Building Cities. City Building according to its Artistic Fundamentals*, Reinhold Publishing Corporation, New York 1945

Raymond Unwin: *Town Planning in Practice: An Introduction to the Art of Designing Cities and Suburbs*, T. F. Unwin, London 1909

Robert Venturi, Denise Scott Brown, Steven Izenour: *Learning from Las Vegas*, MIT Press, Cambridge (Mass.) 1972

THE AUTHORS

Thorsten Bürklin, Ph.D. in philosophy, M.S. in engineering, associate lecturer of urban planning and building science at the University of Applied Sciences in Frankfurt am Main, freelance architect in Karlsruhe.

Michael Peterek, Ph.D. in engineering, professor in the Department of Urban Planning and Design at the University of Applied Sciences in Frankfurt am Main, freelance urban planner in Frankfurt am Main.

导言：从单栋房屋到城市街区

城市并不只是许多单栋房屋的集合，也不只是一座"大建筑"。作为我们日常生活的舞台，城市中的邻里和街区是由一些建筑要素构成的，其尺度介于那些个别的建筑单体与那些被称作邻里甚至整个城区的更大单元之间。这些要素在一栋房屋或者一块用地的个体性（和私密性）与更复杂城市环境的群体性（和公共性）之间起着中介作用。

这些构成要素也可以被称为"城市街区"。它们以不同的形态和几何特征出现在城市的布局中：联排式、街坊式、庭院式、街廊式、行列式、独立式、组团式和"棚厦式"。当然，类型之间也存在着各种组合的可能，而且事实上，各种组合类型都可以在城市现实中找得到。城市街区凭借其特定的形态和独特的组合，通过促进或者限制某种功能而影响着我们的生活方式。因此，对城市街区有所认识也就成了城市设计的一项基本技巧。为了评估设计的效果，城市规划师和建筑师必须掌握城市街区的知识。只有掌握了这些在形态、功能、尺度和重要性方面各不相同的城市要素，才能做好城市设计。

以下各章节从不同的角度呈现了各种类型的城市街区，主要关注城市街区的空间构成、功能、与城市环境的关系、私密与公共领域的关系以及这些类型在城市中出现的条件。各章节还论及了城市中这些基本的结构性要素的演变，并且结合一些历史上或当代的案例来进一步说明。

每一种城市街区的类型都从以下四个方面进行了讨论：
— 形态与空间结构（城市要素的实体描述），
— 城市空间构成（"城市街区"对周边城市空间的影响及其重要性），
— 功能、朝向和连接方式，
— 历史案例。

当然，在现实的建筑与城市中，各种城市街区类型之间的差别并不总是像本书所设定的主题性结构那样清晰可辨，有许多类型是似是而非的、模棱两可的、不伦不类的，很难明确地归入某一类型。

尽管如此，先研究典型的城市街区对于学生而言仍是很重要的，这样可以使他们利用这些知识去分析城市中发现的不同组合与混杂形态，并在自己的设计中有所考虑。出于这样的目的，本书旨在提供一些关于城市街区的常规的、实用的、系统的信息和知识。

P13　**联排式**

联排式街区是一种构成城市与住区的最古老、也是最重要的结构性要素。单栋房屋和单块用地沿着直线、折线或者曲线连接排列,形成街道并通过街道相联系。联排式街区使单栋房屋扩展为更大的城市肌理,是我们的城市与乡村中最大量的也是最基本的空间构成要素。

P13
与街道的关系
扩展的形态

形态与空间结构

联排式街区的构成规则是,建筑的入口及连接通道面向街道,从而在空间上和功能上限定了街道(图1)。

除线性扩展的原则之外,联排式街区可以有完全不同的形态。它可以是开敞的,也可以是封闭的(即联排式);可以是单面的,也可以是双面的。在由独栋住宅或并联住宅组成的开敞的联排式街区中,房屋的周围环绕着开敞空间。对于独栋住宅,开敞空间环绕在四周;而并联住宅则两两成对。在联排住宅中,房屋之间没有缝隙,视觉上形成连续的空间界面。

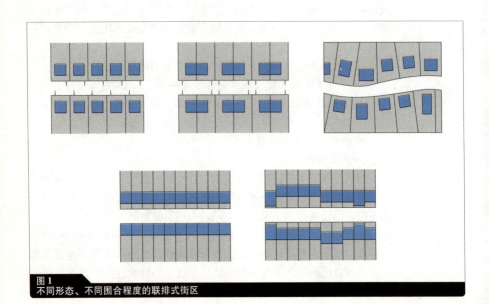

图1
不同形态、不同围合程度的联排式街区

图2
英格兰南部城市巴斯的处于坡地上的单侧联排式街区

单侧的联排式街区（也可以有开敞式和封闭式之分）的建筑只出现在街道一侧；而双侧的联排式街区则在街道两侧都有建筑，但两侧的建筑不一定是相同的。联排街区两侧的建筑样式是各自独立的。联排式街区，尤其是开敞式的联排式街区，可以很好地适应地形的变化（图2）。

统一中的
多样性

联排式街区对于各种形态原则有很好的适应性。在联排式街区中，每栋建筑的外观、三维形体（面宽、进深和高度）以及功能可以是相似的，甚至可以是完全相同的（图3）。它们也可以看上去完全不同，形态各异（图4）。这意味着每栋建筑都可以有不同的外观和特点。

尽管如此，城市中的联排街区个体建筑之间的关系通常是协调的，其基本原因是出于经济上的考虑：相同房屋原型的重复应用有利于快速而廉价地建造。独户、双拼或联排式住宅就是因此而产生的——每栋房屋都具有相同的尺寸、相同的内部与外部形式——并且构成了整个区域的特征。

理论上说，联排式可以在长度上无限延伸。然而，城市基础设施的供给能力及距离长度却是有限的。长度增加可能导致建造不经济。而且，街区过长也会使得建筑后部的用地难以开发利用。在各种街区形态中，为了形成更为经济的城市单元，必须引入垂直的街道将联排街区打断（参见"街坊式"一章）。

出于经济上的考虑，单侧的联排式街区在城市规划中并不常见。而双侧的联排式街区，花费同样的投入，就会使建筑有双倍的机会与

77

图3 皇家广场上的整齐划一的联排式住宅，英国巴斯

图4 阿姆斯特丹码头区的各式各样的联排式住宅

城市基础设施相连接。单侧的联排通常用于某种特殊情况，如住宅区的边缘或与公园、河道相邻等，更容易创造出高质量的居住和工作环境。

P15　　　**城市空间构成**

联排式由于和街道直接联系，从而形成了清晰可辨的城市空间。同时，它们与整个城市的基础设施网络融为一体，进而成为城市空间系统的有机组成部分。联排式可以很灵活地填充城市结构中的缝隙和中介空间，也比较容易跟其他类型的城市构成元素相协调，如街坊式、行列式和独立式等。

前面与背面　　由于联排式直接面向街道的特点，建筑前后的空间就会有明显的差异。这种差异不仅仅表现在功能层面（沿街一侧为公共空间，背后的花园或庭院供私人或集体使用），其建筑的设计特征也是不同的。在街道一侧，建筑设计得既严肃又庄重，而在背面一侧的设计则比较放松，更多受到个体趣味和使用需求差异的影响（如通过改造、扩建增加阳台、檐廊、温室和屋顶花园等）。

向街道空间过渡　　面向街道部分的设计对于城市空间的形成是非常重要的，特别是对于从建筑的私密空间向城市公共空间过渡更是如此。由于区位、朝向、地形、建筑类型等因素的不同，这个前部的过渡区域的设计可以是多种多样的。在高密度的历史城市中心区，为了节省空间，建筑通常与街道直接相连，用地红线和建筑控制线是重合的。这种做法同样

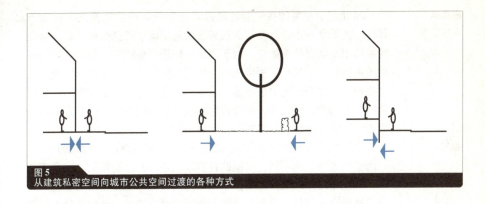

图5
从建筑私密空间向城市公共空间过渡的各种方式

适用于建筑底层为商业用房的情况。但是如果建筑地面层标高与街道地面相同时,对于居住建筑就不合适了,因为那样行人能够窥视私密空间。因此,在当代许多城市的居住区中,建筑师通常会在建筑的私密领域和公共空间之间设置一定距离的缓冲区域。这一缓冲区域可以是一些植物、一个花园,或者是一个私用的前院,为住宅提供一些重要的辅助空间(停车场、车库、自行车存放处、垃圾筒、休闲活动场地等)。即使是一排树也可以看做过渡地带。抬高地面层室内地坪,不但可以避免行人对居室的视线干扰,还可以为住户提供从室内观看街道空间的良好的视点(图5)。

P17
功能混合

功能、朝向和连接方式

联排式和城市街道网络以及城市基础设施有直接的联系,因此它能适应各种城市功能。即便如今,我们依然能够在城市历史中心区的街道上以及在有机演变的住区中找到混合使用的生动案例。联排式街区的地面层空间尤其适合那些非居住的功能(商店、餐饮、小型作坊等)。如有需要,也可以通过扩建或加建将这些功能延伸到用地的后部——在现实中,封闭的联排式街坊的扩展没有多少余地。穿过前排的建筑设置入口有助于将空间向用地后部延伸,为人和车辆提供了更好的可达性。但汽车的驶入、噪声、更多人的进入也会使原本安静的、远离街道的私密领域遭到破坏。

这就是说,联排式街区基本上是作为一种单一功能的城市元素,大多数情况下用于居住。这种做法符合现代城市规划的理念,即《雅典宪章》中所倡导的,将城市功能划分为居住、工作和交通等。

密集的联排住宅

20世纪，由紧挨在一起的联排式住宅所构成的联排式街区大行其道，主要是由于这种经济的布局方式对于城市扩张造成的日益紧张的城市用地状况颇为适应。由于用地可以进一步细分，与那些随处可见的由独栋或双拼住宅所构成的开放式街区相比，联排式街区使得私有土地上的独户住宅单独建造，从而形成高密度。除了住宅构件标准化所带来的经济性之外，联排式街区在节约空间与节省造价方面也有优势，因为个体住宅可以建造在非常小的场地上。

朝向

由于联排式街区沿着街道布置，其建筑朝向取决于街道的走向。而由于用地方位的不同，房屋的自然采光条件差异会很大。联排式街区，特别是封闭的联排式街区，一天之内仅在特定时段内能够得到日照，这种由于区位局限所造成的不利因素可以通过在平面设计上采取适当措施而得以弥补（如采取双面朝向）。

联排式住宅还有一种特殊的类型，即一排房屋叠放在另一排之上。这种"叠拼"式住宅由两套两层住宅叠加而成，上面的一套可由单独的室外走廊进入。这种方案可以形成高密度的城市环境，而且可以获得与有单独入口的独户住宅相同的居住质量与环境气氛（图6）。

功能混合的联排式住宅

近来，"联排式住宅"一词用于描述那些布置成排或成街坊的房屋，这些房屋在密集开发的城市中心区将居住和工作的功能结合起来（图7）。"联排式住宅"至少要有3层，有时是4层，这种住宅为附属的商业功能提供了足够的空间（如一间店铺，或位于底层的办公室），或许还有一间分开建的附加居室，一个私用庭院或花园，有些还附设屋顶平台等室外空间。"联排式住宅"这种住宅类型在建筑上带有极大的多样性和个性特征。

提示：

《雅典宪章》于1933年在国际现代建筑协会（CIAM）第四次会议上获得通过。该次会议在1933年7~8月间在从马赛到雅典的"帕特里斯"（Patris）号邮轮上召开。勒·柯布西耶是该宪章最重要的发起者和起草者。

提示：

联排式住宅地块的宽度通常为5.5~6.5m。在某些情况下，地块可以窄到4.5m。如果该地块进深为25~30m，那么地块面积则为150~180m^2。而双拼式和独栋式住宅用地通常会达到300~400m^2，甚至更多。

> **提示：**
> 东西向联排式可以分别在上、下午接受充足的日照，但在中午则无直接日照。在夏季，阳光会使房间过热，因此有必要考虑在建筑上安装遮阳设施。在北半球，南北向的联排住宅可以从南向的阳光中获得益处，特别是在冬季，温暖的阳光可使房间获得舒适的温度。这种朝向的一个额外好处是节能（被动式太阳能利用），然而，它也有一个缺点，即北向的房间全天得不到一点阳光的照射（除了盛夏时分的傍晚），因此不能把儿童房布置在北向。尽管如此，北向房间却适用于做工作室或者某些特定的工作空间，因为北向光线较恒定、均匀，人们工作时可免受光线变化的影响。需要注意的是，本书中所讨论的关于南北向的建筑规则，到了南半球就完全相反了。

P19

历史案例

最近发掘出的古希腊人的定居点表明，在古代城市中，由相同宅基地和房屋组成的联排式街区发挥着重要作用。这种街区模式之所以被广泛采用，其原因是它为城市用地的划分提供了简单而理性的原则。另外一个优势在于，这种街区构成模式提供了一种可能，使得所有居民受到平等对待（所有人的居住条件和功能都是相同的）。

中世纪的联排住宅

中世纪城市同样基于宅基地和房屋联排布置的构成原理。尽管这些房屋在类型学的意义上是相同的，其建筑细部通常却大相径庭。手艺人和商人的住宅往往带有混合功能，它们构成了城市的基本单元。其前部与城市空间相结合——小巷、街道、广场，而其后部则完全是

图6
伦敦马尔吉斯路居住区的高密度叠拼联排式住宅

图7
卡尔斯鲁厄市中心的联排式住宅

图8
格但斯克老城区由中世纪商人住宅构成的联排式街区

由庭院和花园构成的私人领域，与外界几乎隔绝。随着城市空间的高密度开发，如果需要更多的空间，后部也可以增建空间。典型的例子可以从许多中世纪城镇中找见，如格但斯克、吕贝克和阿姆斯特丹。直到如今，这些城市依然保持着原有的城市气氛和生活质量（图8）。

田园城市

历史城市，特别是19世纪的工业城市，因其过密发展而造成的生活条件恶化而广受诟病。针对这种情况，发源于英国20世纪初的田园城市运动试图创造一种以开敞的或封闭的联排式构成的新型居住区和城市发展方案。这种城市规划与空间构成的目标是创造一个多样化的开放的城市发展结构，以最终实现全面性的包括社会、经济、健康和卫生等方面的改革。城市规划师特别热衷于为居住条件不佳的工人阶级提供绿地里的住所，联排住宅后部的花园可以用于耕种，以满足家庭的日常需求（图9）。

提示：

田园城市的中心著作是1898年出版的埃本尼泽·霍华德的《明日：一条通向真正改革的和平道路》。霍华德希望田园城市能够同时结合城市和乡村的优点。因为这些城市被设计为居住和工作的场所，所以同时包含了工业和文化设施。根据霍华德的观念，六个大型的自治城（每个容纳32000居民）组合在一个中心的城市区域（容纳58000居民）周围。然而，只有几个自治的田园城市被建成。第一个是1904年伦敦北部的莱切沃思。很多被规划为已有城市周边的田园郊区，在此基础之上，保留了功能和经济上的独立。

图9 卡尔斯鲁厄花园城郊的联排住宅

图10 勒默施塔特小区联排住宅，法兰克福

现代住宅区

在20世纪初的新住宅运动中，联排式居住区的设计概念仍大行其道。在布鲁诺·陶特和马丁·瓦格纳设计的柏林住宅区以及由厄恩斯特·梅（时任美因河畔法兰克福市规划局长）设计的勒默施塔特小区（1927~1928年）项目中均可以看到。该小区的设计就是根据这个原理，将标准化的平面布局紧凑的住宅联排布置，每户有自己的花园。这种布局创造出了一种令人愉悦的简单明确的城市街道空间（图10）。

最近一个时期联排式街区的复兴表明，这种街区构成模式的魅力仍有增无减。这是一种有着巨大个性化可能的城市住宅的形态，与多样化的生活方式相适应。同时，大量建造独户住宅造成土地的过量消耗，从而使我们城市外围地带的发展失去控制。寻找像联排住宅那样占地较小的可持续的住宅模式成为当务之急。

> 提示：
> 新住宅运动最早出现于包豪斯时期，这所学校于1919年在魏玛成立，带有手工艺、建筑和视觉艺术工作室。该运动的基本目的是突破历史局限，利用工业化生产方式创建理性的建筑学。

街坊式

和联排式一样，街坊式也是最古老的、最重要的城市设计要素。从古代以来，这种街坊式就对欧洲城市的结构发挥着重大影响。然而，20世纪初期，城市规划师们曾对这种街坊式的采用造成的居住条件的不均提出了质疑。直到20世纪末，街坊作为一种城市要素的积极意义才被重新发现。

形态与空间结构

外部与内部

街坊通常由一组宅基地（在某种特殊情形下也可以是一整块宅基地）组成，其外围各边被街道围绕，并从街道进入。构成街坊的建筑的正立面朝向街道，在街坊内外空间之间形成一个清晰的分界线，且建筑前部是公共领域，后部是私密领域，有很强的朝向感。街坊内部可以保持开敞，也可以部分地或全部地被建筑填满。街坊内部可用作花园、庭院、空地、车库、储藏空间、辅助建筑等。

街坊的几何形状

街坊可以有各种各样的几何形状，可以是三角形、矩形、方形、多边形、椭圆形、半圆形甚至圆形。重要的是，街坊的各个边必须能从外围进入，并面向外围区域。尽管如此，街坊的基本几何形状决定了建筑与城市设计的基本构架（如锐角的转角），决定了内部空间的品质，也决定了住房居室的采光条件。

街坊式建筑的各条边可以是连续的，也可以有些缺口。一种开敞的城市街坊可以由一些较短的联排式、并联式或者独户住宅构成，但房屋之间的距离必须足够靠近，以免有损街坊的整体效果（图11）。

转角的设计

对街坊转角的设计是比较有难度的，这并不完全是从建筑学的角度来看。从一方面看，转角占据位置上的优势（适于开设商店、餐馆或其他商业设施），因为它可以从两个方向接近；从另一方面看，转角也有很多劣势：它的后部用地狭小，甚至根本没有后部用地，转角不适合私用，也不适合扩建，如果建筑朝向不利的话，其过窄的背立面也不利采光。

城市街坊的转角可以设计成缺口，以改善转角建筑的采光条件；也可以完全去掉转角单元或者抹成斜角；还可以把转角建筑加宽或缩窄（图12）。

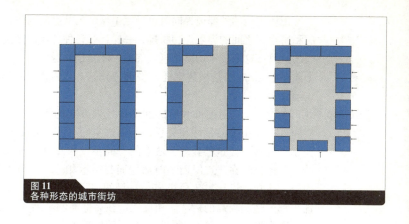

图 11
各种形态的城市街坊

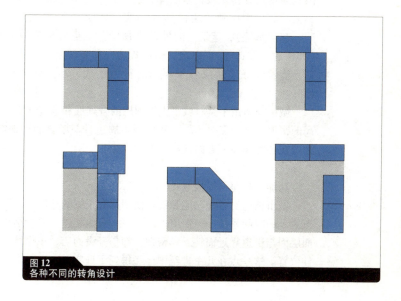

图 12
各种不同的转角设计

P24
整合到
城市之中

城市空间构成

街坊式容易和周围的城市结构紧密地结合。它与城市街道的网格、与建筑控制线相关联，这些建筑控制线在空间和几何形状上限定了街坊。城市街坊是一种界面连续且闭合的城市空间，它从每个边都能进入，从而确保周围的城市肌理和外部的城市空间能够连续。

街坊用地的外部边界同时也确定了城市公共空间与建筑和宅基地的私密空间之间的界限。和联排式一样,有各种各样的方法来设计空间的过渡:这取决于房屋是直接临街,还是稍微退后一些(如设置一个前花园),底层是做住宅还是做商业用房,以及有没有地下室(见"联排式"一章)。

前部与后部

在设计中,对前部和后部的设计处理方式的不同,反映了外部空间(与城市公共领域相关联)与内部空间(与共享的私用空间相关联)的区分。这种处理手法上的不同不仅仅反映在开放空间的设计上,也反映在建筑层面。前部立面与街道相连,是路人看得见的部分,通常要设计得比较讲究,材料的选择、门窗的比例、细部装饰等都要达到较高的标准。相反地,在后部能够看到的人相当有限,因此通常设计得比较讲求实际,窗户的位置不怎么遵从构图的原则,其大小和比例反映了不同房间的实际需求(厨房、卫生间、储藏间、起居室等),建筑上的处理更加随意,更容易适应需求的改变(如扩建、改造等)。总之,街坊式是一种非常复杂而灵活的空间体系,能够较好地适应各种不同的行为模式、不同的活动和不同的趣味。

高密度

城市街坊由于其对城市土地利用的合理性和经济性而可以获得较高的密度。这一点对于眼下我们城市周边日益增长的土地需求来说,无疑在环境和经济方面具有较大的优势。

P25
功能混合

功能、朝向与连接方式

城市街坊对各种功能和混合使用都有很好的适应性,因为它与城市更大的空间系统——街道、广场——直接相连。尽管街坊建筑的地面层因与街道靠近而缺乏私密性,却是开设商店、小作坊和餐馆的绝佳位置。这一点在几个世纪中已经得到了证实。

> 提示:
> 20世纪的城市街坊设计上更追求整体性,内外立面处理得很相似,成为一种新的建筑与城市空间理念,从而也形成了不同的室内外空间特征。

图 13
街坊内部的不同利用方式

城市街坊灵活的后部区域可以为各种各样的活动和使用提供空间，并且其建筑学特征在辅助建筑上亦有所表现。在中世纪和德国的"经济繁荣时期"（gründerzeit）街坊内部往往有小作坊，工作和居住混在一起，我们也可以看到整个工厂建筑坐落于街坊内部的情况（图13）。如果需要的话，车道可穿过前部的建筑抵达街坊后院。

20世纪早期出现了一场运动，禁止那些有噪声和污染干扰的企业设在城市街坊内部。这一运动和《雅典宪章》有关。该宪章提倡对城市功能进行明确的划分，如居住、工作、休憩和交通（参见"联排式"一章）。从那时起，街坊式布局主要用于居住区，而街坊内部的区域则用于私用的或半私用的活动场地、开放空间、花园、绿地等。

直到20世纪70年代，与无干扰的功能进行平衡组合的做法才再一次回到城市之中。由工业向服务业的转型对工作场所及其可能造成的干扰有所改变，使之更容易与周围的居住区相结合。在多数情况下，功能的混合不但不是问题，而且能创造一种特殊的品质。城市街坊得以继续为功能混合提供出色的条件，即使达不到所有街坊作为商业用途时那样的密度。在这种情况下，沿主干道的密度往往可以达到最大，而附近的建筑底层空间往往作用于居住目的。

建筑的进深与朝向

由于城市街坊的边界遵循街道的走向，那么房屋的朝向就会存在某种限制。在有双朝向的住宅中，辅助性房间可以布置在缺乏自然光照的内院一侧（这里假定东西向房屋的进深平均为 11～13m）。南北向住宅的平面要设计得宽大一点，进深也要小一些，大约 9～11m 为宜。这样有利于充分利用南向立面吸收直射阳光。

除了日照方面，街坊式住宅的朝向还要考虑街道交通以及由此带来的噪声干扰。在很多情况下，建筑师要在朝向太阳（但要暴露在街道噪声之中）或是朝向安静的后院（却要面向背阴之侧）的矛盾之间权衡利弊。在这里，我们建议在进行平面设计时采用双面朝向以满足各种需求。

停车空间

停车空间往往布置在街坊前面，与街道平行、成夹角或垂直。如果每 5～10 个车位用树木有规律地间隔一下，就会形成一种更动人的绿色城市街景。由于现代城市交通量的骤增，停车空间很难满足需求。在这种情况下，在房屋或内院下方设置地下车库也许是十分必要的。然而，设置地下车库可能会给院内植物的种植带来制约，而且会增加造价。重要的一点是，不要在街坊内部布置地面停车场，否则不但会削弱内院的视觉效果，还会带来噪声，破坏后院的宁静气氛。

P27
古代的城市街坊

历史案例

从古至今，城市街坊一直充当着最重要的城市设计要素。它最早用于公元前 6 世纪的希腊城市，公元前 5 世纪希波达姆斯设计的米利

注释：
　　在混合使用中，为确保灵活性，街坊底层的层高可以略微增加一些，特别是沿街的部分提高到 3.0～3.25m，而不是居住功能所需的 2.5m。

注释：
　　首要的规则是，如果一幢街坊建筑只有 3 层高，就可以假定沿街道连续斜向停车，可以满足每户一辆车的车位需求。然而，如果建筑层数超过 3 层，则应采取其他方法停车。

图14
希腊城市米利都城平面图

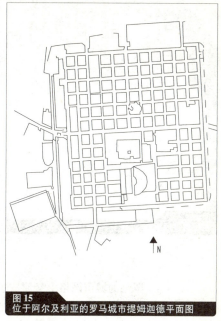

图15
位于阿尔及利亚的罗马城市提姆迦德平面图

都城最近被发现，该城的规划即是以规则的正交网格为基础的（图14）。许多希腊人的定居点都是以街道系统为布局基础的，如奥林图斯（Olynthus）、阿格里真托（Agrigento）、帕埃斯图姆（Paestum）和奈阿波利斯（Neapolis）等。

罗马的城镇规划继承了网格的原理，并更加严格地加以应用在其新建的城镇中，如科隆、特里尔、尼姆、博洛尼亚和佛罗伦萨等。这些定居点往往是从兵营（castrum）演化而来的，其基本骨架是两条垂直相交的主干道——南北向的大道（cardo maximus）和东西向的大道（decumanus maximus）。这两条轴线把城市划分为四个区域（图15），因此至今我们仍把"城区"称为"四分之一"（city quarter）。市场和重要公共建筑布置在这两条主干道的交叉点附近，而其余的次要街道与主干道系统平行布置，形成街坊结构。由于城市特殊的地形特征（如山丘、河流等）以及与原有街道的结合，这个网格系统会产生变形，困难的地理区位可以造成三角形或多边形的街坊形状。

中世纪的城市街坊　在许多地方，罗马城市布局在经历了罗马时期之后的人口减少和城市衰败之后仍然保持下来，直到中世纪再次得到复兴。尽管在

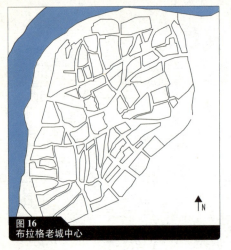

图 16
布拉格老城中心

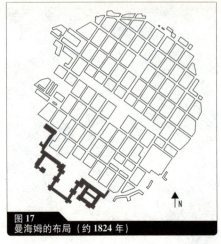

图 17
曼海姆的布局（约 1824 年）

罗马的城市网格上又建造了新的房屋，但街道布局和街坊结构却基本未变。大多数新的中世纪城镇以及扩展的城区没有遵循罗马的网格系统，而是采用了大小不同、形状各异的多边形街坊系统。这样便形成了独具特色的由街道、小巷和广场组成的公共性城市空间，为房屋的进入提供了通道，同时也确保了社会生活与商业活动的需要。与之相对应的是，私密的后部区域由辅助建筑、庭院和花园组成（图16）。

殖民地的城市

文艺复兴时期的新兴城市（如建于 1593 年的位于威尼斯东北部的要塞城市新帕尔马）和巴洛克时期的城市规划（如 17 世纪的曼海姆）从古代继承了规则的格网式布局（图17），北美和南美的一些新兴城市也是如此。西班牙和葡萄牙的占领者向新世界输入了规则街坊的城市布局理念，作为城市设计的正式原则。如今，这些基本图案作为核心组织结构在某些城市中依然存在，如墨西哥城、利马、加拉加斯和圣多明各。北美最著名的例子要数曼哈顿。这座由荷兰移民建造的城市采用棋盘式的图案作为城市布局的基础，而如今带有高层建筑和摩天大楼的城市景观和传统的城市街坊已经不可同日而语，那时的房屋没有这么高。

工业时代的城市

19 世纪的工业时代快速增长的城市也采用了街坊结构，因为这种结构有许多优点：易于将城市整合为一体，对各种功能有很大的包容性，建筑密度和人口密度高。

图18
位于柏林普伦兹劳贝格地区的"经济繁荣时期"的街坊

在德国"经济繁荣时期"（1871～1914年），柏林扩张期间，城市规划师们开始建造一种高密度的城市街坊，这种街坊带有多重内院，并设有车道与街道相连。这种街坊占据很小的街道面宽，却有着很大的用地进深和很高的密度，非常适合当时用地紧张的状况。然而，正由于这种高密度的开发，后院住户的许多房间（通常住有15人之多）得不到直接日照，甚至连采光也不够。由于密度达到了每公顷居住人数超过1000人，大多数情况下卫生条件极差，生活条件简直糟透了。肺结核及其他传染性疾病到处蔓延（图18）。

早在19世纪初期，这种糟糕的社会条件和卫生状况遭到了许多人的尖锐批评，包括弗雷德里希·恩格斯。他在1845年写下了《英国工人阶级的状况》一书。到了20世纪初，这些批评引发了对城市街坊的一个局部的改革。在"现代"的街坊中，曾经占据内院的那些房屋被绿地取代了。H·P·贝尔拉格在阿姆斯特丹、J·J·P·奥德在鹿特丹，以及弗里茨·舒马赫在汉堡的方案中，都可以看到这种改革的做法。

在20世纪20年代，"新住宅"运动（参见"联排式"一章）的代表人物纷纷采用独立的行列式作为城市的主要构成要素来替换封闭的街坊（参见"行列式"一章）。这一做法从根本上改变了欧洲城市的外观形态，此后的几十年，作为城市元素的城市街坊变得不那么重要了。

图19
柏林腓特烈城区南部的新的城市街坊

图20
腓特烈城区南部的街坊内部安静的共享区域

城市街坊的复兴

从20世纪60年代开始，直至七八十年代，城市街坊在法国、意大利、德国以及其他欧洲国家卷土重来。起因正是对现代建筑与"新住宅"运动对城市造成的破坏性影响的批判。1980~1990年间在柏林国际住宅展（IBA）上建造的住宅项目体现了这种理念的转变，也引领了"城市重建"运动（图19、图20）。

如今，城市街坊再次成为城市规划的重要工具，甚至那些几年前还饱受诟病的德国"经济繁荣时期"建造的街区，也因其高密度和多功能而再受追捧。后院的建筑即使它们密匝匝地挤在一起也不再被清除干净。而且这些房屋的所有者往往会对这些房子进行改造，改变功能，做工作室、画室、无干扰的小作坊，使之更有魅力，更与众不同。如果街坊内部空间足够大的话，有时甚至会增建新建筑。

提示：
1966年，建筑师和理论家阿尔多·罗西出版了有影响力的著作《城市建筑学》。在该著作中，罗西强调了街坊结构对于形成城市空间的重要作用。他还强调了这种城市结构的一致性（稳定性）及其对社会、社会认同感以及历史价值的重要性。

提示：
19世纪工业城市的高密度的街坊结构重获青睐的原因是，人口密度下降而建筑密度保持不变。如今两个人住在三居室的公寓中并不鲜见，而一个世纪以前，这套公寓要容纳25~30个人。

庭院式（反转的街坊式）

从城市组织的意义上说，庭院式可以看做反转的街坊式。在街坊式和庭院式中，房屋布局是相同的。但是在街坊中的建筑（即那些沿街坊周边布置的房屋）是从外部进入的；而在庭院式布局中，是从内部进入的。因此，庭院式布局中房屋的正面朝向内部空间，而背朝外部。内部区域成了（至少一部分成了）公共空间。

作为城市规划术语，"庭院"一词来源于围合的农场场院或修道院之类的原型。在这种原型中，房屋围绕一个院子成组布置。庭院这个词因此被用来表示一种以一个开敞空间作为形态和功能组织中心的一组建筑。作为一个整体，这组建筑带有自治的、内向的气质。

形态与空间结构

庭院式通常作为一个完整的单元，其布局在很大程度上是以邻里和集体生活为基础的（图21）。

庭院可以被一些相似的建筑围合，或者它们可以被一组形态各异的建筑所围合。无论上述哪种情况，重要的一点是，庭院边界要尽可能地在空间上闭合。除了人和车的入口以外，不能再有其他开口打破庭院的围合感。如果建筑物本身不能形成边界，就需要制造其他能够形成边界的要素，如围墙、绿篱等。

和城市街坊一样，庭院也可以有完全不同的几何形态。比如，作为前院或入口庭院，其功能也相当于街坊结构的一个次级要素。

由于庭院建筑的正面朝内，而其朝外的一面与公共空间相邻，因此其两侧的立面必须满足不同的设计要求。和城市街坊相反，庭院内部的形态设计要有严格的控制，不允许自由随意地改扩建。在庭院中，立面的设计和材料的使用前后或内外的差异并不是很大。

> **提示：**
> 围合（close）一词在英语国家里也有"庭院"之意。该词源自拉丁语 claustum，意指修道院。德语中的 klaus 也是这么来的（一幢建筑或一组建筑围合起来，与外界隔绝；修道院）。

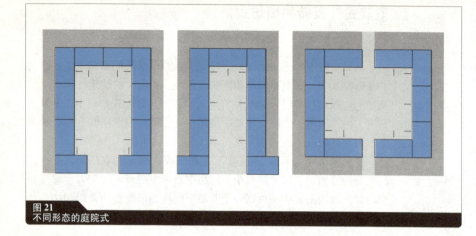

图 21
不同形态的庭院式

P34
半开放空间

城市空间构成

庭院与城市公共街道系统是分离的。当它与公共空间相连接时（否则建筑无法进入），就形成了一种带有有限的公共性特征的空间，我们可以称之为"半公共空间"。从城市空间向庭院过渡的设计是非常重要的，可以通过坡道和台阶、入口车道形成高差，使空间有层次感；还可以通过不同的铺地、植被和其他手法来达到这一目的。

庭院不像街坊那样与城市环境交织得那么紧密（参见"街坊式"一章）。为了追求内向性，庭院的入口往往采用尽端式道路，故意不去延续城市的网络。庭院自身保持一个自成一体的小天地。如果利用庭院的空间序列最终与公共的街道空间相连接，这种城市要素则可以更好地为其所在的环境相融合。庭院式也就变成了所谓的"街廊式"（参见"街廊式"一章）。

P35
集体使用

功能、朝向和连接方式

庭院式常常作为一种城市设计模型，用于集体化（或合作化）住宅。这种布局为居住者提供了一个定位参照点，一个中心，形成某种带有私密性的安静空间，从熙熙攘攘的城市环境中分离出来。庭院式在一片邻里环境中形成自成一体的单元，这有利于加强居民的安全感，提高他们对集体空间的监控能力，因为住在同一个庭院的居民互相认识，若有陌生人进入会很快被识别出来。通过把重要的元素朝向外部（如通道开放空间或公用区域），建筑师可以强化庭院作为社会空间的诉求。功能混合也是可能的。某些特殊功能和无干扰的服务业如办公、诊所等，可以结合到这种城市元素之中。

庭院和街坊一样面临朝向与日照的问题。而且同样，转角部位的设计也是多种多样的（参见"街坊式"一章）。阴角的设计难度在于更小的建筑面宽朝向院子，要与花园或后部大的室外区域相呼应。

高密度开发

庭院式可使城市用地得到最优化利用。和街坊式一样，庭院式常常用于提高建筑密度。由于它是由内部进入的，可以使最内部的用地得到利用。

除了能提供连接通道的功能，庭院的内部空间可以作为共享交流的场所，儿童游戏场、聚会地点、自行车和婴儿车存放处、垃圾转运站、休闲小公园等。停车场尽可能不要布置在内院中，以免破坏其休闲的品质和居住的安宁。停车场应当布置在庭院之外或者庭院之下的地下车库内（参见"街坊式"一章）。

P36
庭院住宅

历史案例

从古至今都有房屋围绕着一个或几个院子布置的做法，尤其是在地中海地区。这种做法在伊斯兰建筑中更为普遍。然而，从城市规划的视角来看，那时的庭院住宅还不算典型的城市建筑类型，因为它通常是建在一块独立的用地上。

农场用房和修道院

我们还可以举出另外一个历史案例，那些在许多地方都能见到的农场用房和修道院，都是从外部环境中围合出一块空间来。修道院和农场用房二者的共同之处在于，它们都并非为单一的居住功能而建。二者的重要特性是，它们利用庭院形成了对外部的防御，同时也形成了受保护的社会空间。意大利北部帕维亚的切尔托萨修道院为了给修士们提供住所而扩建，就是采用的这种围合式的庭院结构（图22）。

慈善住宅项目

雅各布·福格尔在1520年左右为奥格斯堡的穷人所建的居住综合体可以看做现代之前公社大院的范例，其设计原型是中世纪荷兰城市中的女子修道院（hofjes）。这些慈善设施最早可以追溯到13世纪，常常被设定为一种基金会，为有需要的社会群体提供住房，包括老人、穷人、病人和孤儿。阿姆斯特丹的女修道院（bengijnhof）是其中最著名的一个例子。

庭院式街区的概念不止一次地被社会运动所采纳，因为它不但能保障一定数量的共享开放空间，还在高密度的情况下提供了一定的私密性。出于同样的原因，19世纪的工业家们采用这种城市建筑街区结构作为一种工人住宅原型。甚至20世纪20年代广泛流行的"维也纳公社大楼"（Wiener Gemindebauten）也采用了庭院式，米歇尔·布林克曼在鹿特丹—斯旁根地区设计的能容纳270户居住的大型庭院住宅（建于1910~1922年间）也采用了这一概念。在该项目中，除了地面层设置了入口外，二层的室外走廊上也设置了入口。(图23)

注释：
　　如果需要设置地下车库，可以考虑把庭院地坪抬高1m。这种做法可以缩短车库入口处的坡道，而且可以提供组织自然通风的机会。

提示：
　　"维也纳公社大楼"是"红色维也纳"时期的社会民主党政府为解决工人住房问题而采取的措施。根据这项1923年颁布的建设计划，每年要建3万套住房。"维也纳大院"随之出现了。这种住宅具有纪念性的尺度，层高较大，有共享的院子和许多配套设施。其中最著名的要数卡尔·马克思大院，它包含1300个住宅单元和许多商业服务设施。

图22
帕维亚修道院的庭院

图23
建于鹿特丹—斯旁根地区的外廊式住宅庭院（1919～1922年）。米歇尔·布林克曼设计

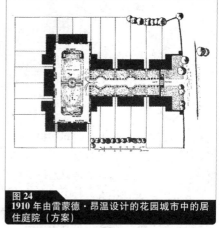

图24
1910年由雷蒙德·昂温设计的花园城市中的居住庭院（方案）

花园城市中的庭院式

　　20世纪初的花园城市运动也采用了由联排住宅围成庭院的布局模式，在绿地上形成安静的住宅组团。在这里，庭院被称为"close"（围院）。邻里关系和小镇的认同感得到格外重视。在英格兰南部莱切沃思、威尔温和汉普斯泰德等花园城市中，建筑师雷蒙德·昂温设计了许多"围院"住宅的佳例（图24）。

　　庭院式在公共住宅特别是实验性或合作化住宅的设计中依然有广泛的应用前景。

> **提示：**
> 1910年雷蒙德·昂温在他出版的著作《城镇规划实务》中描述了新住宅小区中居住性庭院的功能特征和设计要点。他还提到了庭院式住宅的一些优点，如用地经济，住宅因围绕绿地和开放空间布置而视野开阔等。

街廊式

街廊式从有顶的街道演变而来,这种街道两侧排列着店铺,一个紧挨一个。街廊和庭院式在空间结构上有一定关系,它们都是从内部进入的。

P39
带玻璃顶的街道

形态与空间结构

大多数情况下,街廊是一条被玻璃顶覆盖的购物商业街。作为公共通道,通常只允许步行者进入。其两侧界面均由紧挨在一起的建筑立面围合,每个立面都经过了精心的设计,给人以富丽堂皇的印象。透过巨大的玻璃窗,街上经过的所有人都能看到商店里展示的商品。

街廊可以是直线形的,也可以是折线或曲线的。它可以采取几乎所有可能的线性形态,或者向不同的方向分岔。在街廊的交叉点上空间会有所加宽,从而形成小广场供人们逗留(图25)。

街廊可以穿行于两个不同的建筑物之间(有时甚至是多层的建筑),也可以作为贯穿紧凑的城市街坊的公共通道。在这种情况下,则要格外注意内立面的设计,内立面往往要用心设计,以反映出外立面的形式特征。

P40
通道的网络

城市空间构成

街廊连接着城市中的通道。在巴黎、布鲁塞尔、伦敦、那不勒斯和米兰,有许多著名的19世纪的街廊。例如米兰著名的维多里奥·伊曼努尔二世街廊,就是大教堂和斯卡拉大剧院这两个重要的城市节

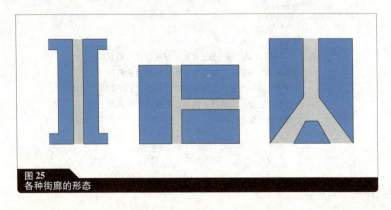

图25
各种街廊的形态

图26
米兰的维多里奥·伊曼努尔二世街廊

点之间的最短路径（图26）。有些时候，至少对步行者而言，街廊可以形成一些捷径，从而对城市的通道网络起到优化的作用。汉堡的有玻璃顶棚的街廊不但有遮风挡雨的功能，同时也为进入市中心提供了一条重要的通道。

气候缓冲区　　由于屋顶的作用，街廊内部形成了气候缓冲区，使街坊作为供人逗留的场所更有吸引力，特别是在那些气候恶劣的季节。如今，许多街廊冬季有采暖，夏季有空调，但这种街廊需要在空间上进行明确的分隔。这种街廊的空间特征更室内化，更像一座百货商场，与城市空间分离，使之看上去不像一条连续的有顶街道。

P41

功能、朝向和连接方式

街廊的设置主要出于对"经济性"的考虑，因为在城市中心区位，街廊可以为街坊内部用地提供进入的机会。街廊的地面通常做得很平，为的是不分散行人的注意力，使他们能够专心地欣赏商店橱窗（图27）。

商业功能　　街廊的功能通常以零售商业为主，偶尔辅以餐馆。住宅更不常

图 27
葡萄牙阿维罗市中心新建的街廊

见。但是如果街廊中有住宅，为了防火的需要，也为了便于组织日照、采光和通风，玻璃顶盖通常要设在住宅层以下。因此可以说，把居住功能结合到街廊中的唯一影响是，住宅入口要设在街廊之内。

街廊要考虑的另一个重要方面，不仅仅在于其商业上的成功，还在于其功能上能够整合城市环境。街廊的前部和后部都是同样重要的。建筑师必须确保把各个入口开设在朝向繁华街道的一侧，因为如果不是这样的话，就会导致建筑前部和后部之间的吸引力失去平衡。

广场与市场

历史案例

古罗马有一种空间结构可以看做街廊的先驱。尤利乌斯广场（Fourm Iulium）和图拉真市场周围都有一些商业，二者之间的街道是人们闲逛和谈生意的场所。波斯城市伊斯法罕也有一种非常相似的空间结构，伊斯兰世界的市场和集市（souks）至今还沿用着这种空间组织方式。商铺沿着中间的通道两侧依次排开，各种商品琳琅满目，令人目不暇接（图28）。

图 28
熙熙攘攘的开罗市场

图 29
建于 19 世纪的法兰克福火车站附近的恺撒街廊

19 世纪的街廊

　　在 19 世纪，街廊在巴黎、米兰、维也纳等欧洲城市颇为流行。这些街廊给富有的中产阶级提供了一个与喧闹、脏乱的街道不同的、免受气候影响的闲逛之所（图 29）。街廊通过设计、组织和展览技巧，制造一些乐趣，以满足客户的消费心理和需求，同时对于富有的中产阶级和城市本身，也有一定的炫耀功能。

购物中心与购物街

　　街廊的空间模式作为商业展示之所，可以看做现代购物中心和大型购物街的前身。与街廊不同的是，现代购物中心并非与周围的城市环境结合为一体。从外部看，这些毫无特点、毫无生气的"方盒子"对于城市空间是消极的（参见"棚厦式"一章）。

提示：

　　有关街廊的最重要的著作要数沃尔特·本雅明的《街廊》（德语版书名：PassagenWerk；英文版书名：The Arcades Project）。该书对 19 世纪街廊从文学和建筑学的角度进行了研究，指出了作为社会空间和商业空间的街廊具有美学的、经济的和实用的价值。本雅明还指出，行人在街廊中闲逛，不但是为了浏览商品，也是为了找乐。

提示：

　　2001 年出版的《城市研究：哈佛设计学院购物指南》是建筑师和理论家雷姆·库哈斯和他带领的一个团队共同撰写的一部研究购物中心和商业街历史发展的著作。书中有大量古代罗马市场、古代波斯与阿拉伯的市场与现代购物中心、商业街的图片。

行列式

行列式（德文：Zeilen）是一种线性的独立的城市元素，故意背离街道以获取更好的"卫生"条件，即最大限度的日照和自然通风。行列式发明于20世纪20年代，是为了应对传统城市中街坊结构和走廊式街道所造成的过度拥挤的问题。因此它也可以被看做对19世纪晚期建造的居住条件恶劣的住房的一种批判（参见"街坊式"一章）。

形态与空间结构

行列式可以看做联排式的进一步发展。然而，与之相反的是，行列式的设计并不是为了形成街道空间的界面。在大多数情况下，只有房屋的"山墙"或短边朝向街道。行列式由于独立于街道布置，从而有条件获得最大限度的日照。

行列式与街道不是平行，而是垂直，并通过次一级的人行步道与街道相连（有时是尽端式的）(图30)。行列式房屋往往从采光条件较差的一面进入，也就是说，从东面或北面进入（在南半球则从南面进入）。这样就避免了采光条件好的南面和西面受到进入通道的干扰。

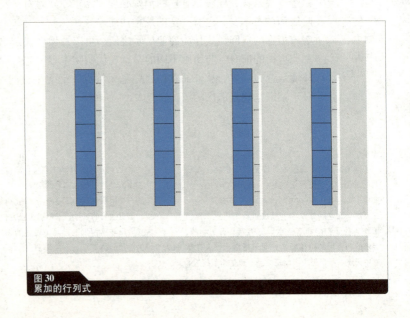

图30
累加的行列式

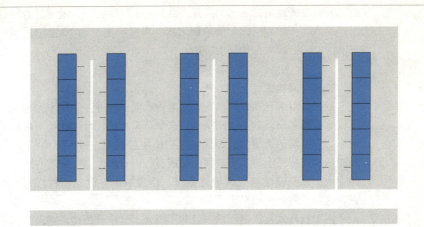

图 31
成对的行列式

这些面常常用于布置私密性的开放空间，如阳台、檐廊、屋顶平台或者位于地面层的小花园。

累加的
行列式

行列式可以通过简单重复累加的方式进行扩展，即一幢的前面对着另一幢的背面。这种直接把后部空间（私密的）和前部空间（公共的）相叠加的方式会使空间缺乏限定感，当然这种缺点也可以通过采取某种措施进行弥补，如通过植物配置、改变层数、增建自行车棚和储物间等，使每幢房屋看起来有所区别。

成对的
行列式

另一种可能的扩展方式是将房屋及其入口进行"镜像"，这种做法会形成一系列成对布置的行列式，面对面、背对背地排列（图31）。这种布局方式意味着，每幢住宅特别是宅间的空间日照条件并不平均，但其外部空间可以获得较好的社会性特征。

构成行列式的可以是连成一线的独户住宅（2~3层的联排式住宅），也可以是集合住宅（3~6层的公寓式住宅），8层以上的板式居住综合体往往也被当作行列式的特殊形式。

大多数行列式住宅是直线的，但也可以是弧线的、折线的或者是锯齿状的。通过房屋长度和高度上的变化，城市空间的形态可以得到塑造。

标准化

行列式可以被理解为大批量生产时代的产物，其线性特征和个体单元的重复性，使之非常适合工业化预制生产的需要。然而这种在建造上具有经济性的标准化生产方式也有一定风险，即大量重复会导致形式的千篇一律和空间上的单调乏味。典型的例子是20世纪后半叶东欧国家建造的大量预制大板楼。

P46　　　　**城市空间构成**

行列式以采光和日照确定朝向的做法使得这种建筑几乎完全独立于城市空间和周围的街道网络。从这个意义上说，行列式是对传统城市形态和空间的否定。这种独立性往往使之成为一种反城市的要素，不能在常规意义上对城市空间的构成有任何贡献。当行列式在市中心的空地上或两幢建筑夹缝中建造时，作为城市界面的结构性构成元素，这种矛盾就更加突出。

流动空间　　由于行列式建筑之间的空地并不围合，这样就使得该空间成了缺乏限定的公私不分的流动空间。这片同质性的用地通常由草坪或其他植被覆盖，其设定的功能是公共交往空间，特别是当行列式建筑为公寓时更是如此。然而，这个空间实际上很少起到这样的作用。相反地，该空间处于匿名状态，没有人感觉与之相关，很快就会被忽视（图32）。更有甚者，由于缺乏有限定感的街道空间，这个宅间空地的社会归属感也是很差的，居民缺乏安全感，特别是那些大型居住区更是如此。

P47　　　　**功能、朝向和连接方式**

居住功能　　作为城市街区，行列式符合功能主义的城市设计理念，正如《雅典宪章》（1933年）中所提倡的，把居住、工作、交通和休憩严格地区分开来（参见"联排式"一章）。由于这个原因，行列式只能用于居住建筑，而用于办公、商业建筑的情况极为少见。由于行列式布局的建筑背离主要街道而很难被路人直接看到，因此并不适合作为公共使用。

正是出于上述原因，一些小型的商业服务设施有时会布置在沿街的行列式建筑的端部（图33）。这些小建筑在城市规划上具有双重的作用。它们不但能在一定程度上修复公共性街道空间的连续性，也能为行列式住宅之间的空间提供某种防护感，遮挡街道上的噪声。其结果使得街道空间和住宅区内的空间都得到限定——这是向塑造城市空间回归的第一步。这样就出现了一种带有安静的半公共空间的介于行列式和（开敞式）街坊之间的形态。

图32
某个现代住宅区中行列式住宅间的消极空间

图33
垂直布置的低层的商业建筑可以从街道一侧围合行列式住宅之间的空间

朝向　　和联排式和街坊式一样，在行列式住宅布局中，东西向和南北向的住宅必须区别对待。在东西向的住宅中，居室可以得到双向的日照，而在南北向的住宅中只能有一侧得到日照。因此，在确定住宅进深尺寸和平面布局时，朝向是要考虑的重要因素（参见"联排式"、"街坊式"章节）。

居住区内的小路　　由于行列式住宅只在端部与城市道路系统相联系，因此住宅单元入口往往需要通过宅前步行小路才能进入。然而在某些情况下，我们也可以看到另一种道路设计模式，即在行列式住宅的一侧设尽端式车道，而在另一侧设置步行小道。这种做法进而又会对小区内部的布局方式、主入口的位置、公寓内居室的朝向等造成影响。这种交通组织形式的优点是人车分流。连接行列式住宅的步行小道往往带有半公共特征，在单元入口附近，鼓励休闲和邻里交往、儿童游戏。

外部空间　　长期以来，居住在地面层的居民并不能利用其宅前的外部空间，因其室内地坪略高于室外地坪。把宅前绿地据为己有被认为是违背了"人人平等"的原则。只有到了最近，人们才认识到，其实居民并不一定有相同的兴趣，一些人喜欢有个小花园，而另一些人则喜欢有个阳台或者屋顶平台。而且，允许地面层的住宅拥有一个小花园或者小院，不但可以使外部空间的美学质量得到改善，还可以加强居民的归属感，提升社会监视感。这样有利于整个居住区的安全性。

历史案例

行列式相对来说是一种新的街区构成类型。历史上曾经有少量的例子可以看做行列式的先驱，如建造于 1768～1772 年间由建筑师亚当兄弟（James and Robert Adam）设计的伦敦艾德菲住宅区，以及 19 世纪晚期在意大利北部建造的外廊式住宅等，除此之外，行列式住宅基本上可以看做 20 世纪 20 年代"新住宅"运动的产物。

最著名的例子是达默斯托克住宅区，该住宅区作为一个展览项目于 1927～1928 年间建于卡尔斯鲁厄（图34）。最终版的总平面是由奥托·海斯勒和时任包豪斯校长的格罗皮乌斯共同设计的。这个规划当时是作为一种全新的城市结构的煽动性宣言而推出的。它全部采用东西朝向板式住宅，形成严谨的南北行列式布局，看上去是无休止的线段，相等的间距，完全放弃了传统的空间组织原则。也许没有任何一种城市规划方案像行列式那样从一开始就引起那么大的争议，有人认为这种小区为每个人提供了优越的生活条件（采光、通风、日照），从而推动了现代城市发展的进步；另外一些人则指责它过于死板的规划原则，形成了抽象的总图、单调的建筑、乏味的空间（图35）。

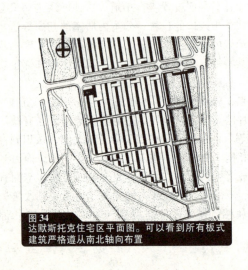

图 34
达默斯托克住宅区平面图。可以看到所有板式建筑严格遵从南北轴向布置

> **提示：**
> 带有外廊式入口的街坊式住宅可以看做由连在一起的住宅单元构成的延长了的行列式，外部通道可以从单层或多层连接起来。

> **提示：**
> 行列式之所以被当作现代主义的代表，一是因为它在城市"卫生"方面带来的进步，二是它所提倡的"平均主义"的社会理想使人人都享有相同的居住和生活条件。同时城市建筑类型对于建筑元素的批量化生产也具有经济上的优势。1930 年，在布鲁塞尔召开的 CIAM 大会"理性的场地规划"单元中，行列式规划理论与手法第一次被系统、全面地讨论。

达默斯托克住宅区所确立的规划模式在 20 世纪 20 年代后期对许多其他著名的"新住宅"运动项目有所启示，如位于法兰克福的海勒霍夫住宅区（1929 ~ 1932 年）和维斯特豪森小区（1929 ~ 1931 年）；位于柏林的西门子城（1929 ~ 1932 年）和哈泽尔霍斯特小区（1928 ~ 1931 年）以及位于卡塞尔的罗滕堡小区（1929 ~ 1931 年）。

图 35
位于卡尔斯鲁厄的达默斯托克住宅区的行列式布局规整而严谨

现代化 　　到了50年代和60年代，这种布局理念不但在欧洲广泛采用，而且成了世界各地低收入阶层的住宅建筑的标准蓝图。
　　20世纪70年代，行列式住宅作为一种功能主义的城市构成元素，遭到后现代主义者的抨击。由于其住户阶层不高所带来的固有的社会问题，功能单一且不完善（被称为"睡城"），以及形态单调，缺乏美感，这种住宅模式很快便陷入声名狼藉的地步。然而，90年代以来，许多行列式住宅区成功地进行了改造和更新（如增加阳台空间、拆除过高的部分、改善周围外部环境等），使其居住条件大为改善。

P51 ## 独栋式

　　在城市规划术语中，独栋式指的是孑然独立或者能够从周围的城市环境中分辨出来的建筑，如农庄、谷仓、城堡、修道院等，它们自古以来就已成为一种文化景观。然而在更为高密度的城市肌理中，独栋式主要是指那些从由联排式和街坊式构成的标准的城市肌理中脱颖而出的建筑。它们通常是公共建筑（神庙、教堂、市政厅）或者是为统治阶级服务的建筑（城堡、要塞）。到后来，它们是为富人建的住宅（别墅、府邸）或者是越来越多的城市基础设施（学校、剧院、博物馆、医院、议会建筑、大学等）。如今在大城市中的独栋式建筑还包括高层住宅和写字楼，以及那些独立式的住宅。这些建筑对用地的需求越来越大。

P51 ### 形态与空间结构
　　独栋式建筑在规模大小、重要程度、几何形状、设计质量和结构材料等方面与周围建筑有很大差异。即使它们在空间上并未跟周围建筑分离，也容易作为一种独善其身的结构性单元。从形态和装饰风格上与众不同，很容易被识别出来（图36）。

形态的自治性 　　独栋式不与任何其他建筑相连，无论从形态还是尺度上，其设计相对独立于周围的城市文脉。这意味着在独栋式的设计中，相对于其他城市街区构成类型，建筑师更有自由度，其形态可以采取板式、塔式、立方体、圆柱体、金字塔形以及各种体量的杂交、组合。然而，当独栋式与更大的城市整体结合时，或者有某种城市轮廓或景观的需要时，其大小、形态和立面也要与某种特定的设计要求相符合。

图 36
各种形态的独栋式

图37
杜加（Dougga）山城的神庙。该古城位于突尼斯，曾经被罗马人占领

P51

城市空间构成

独栋式的设计概念是，不去与周围建筑建立任何直接的关联。在很多场合下，设计的目的是使该建筑明显区别于周围的城市结构，成为城市景观中的一个焦点，形成一种特殊的空间效果。

位置显要　　在某些情况下，独栋式故意选址在显要的区位，与周围的城市结构脱离。古代的神庙和圣地就是如此。这些建筑并未特意塑造城市空间，而是形成了整个城市交响中生动的、雕塑性的高潮部分，使城市交响曲得到共鸣和加强。把这种建筑建在地理上特别重要的位置可以强化这种效果，正如雅典卫城以及许多教堂和宗教建筑等（图37）。在巴洛克和绝对君权时期以及19世纪晚期的城市中，城市规划师们特别重视独栋式的利用，把它们布置在城市主要道路或者视觉轴线尽端或交叉点上（图38）。

融于城市　　然而，在空间局促、建筑密集的城市中，独栋式并非完全独立。它们往往在空间上从属于城市广场的一个界面，或者一排建筑、一组

图38 巴黎的玛德莱纳教堂（Madeleine，又译：抹大拉教堂）坐落于街道与视轴的交叉点上

图39 圣乔万尼和圣保罗教堂（Santi Giovanni e Paulo）融入了威尼斯的城市肌理之中

建筑中的一栋（图39）。这种情况在密集的紧凑的中世纪城市中尤为明显，大教堂、市政厅、修道院甚至什一税仓库（tithe barns，中世纪时欧洲北部的一种仓库，专门用于储存作为"什一税"的农产品——那时的农民必须把十分之一的农产品作为税，上缴教会——译者注）都混迹在城市肌理中，但仍然能保持与众不同，在规模大小、立面风格以及在城市布局中的独特地位上有别于周围环境。

空间效果　　在现代城市中，独栋式建筑往往独立而建，其朝向主要考虑日照和通风而确定（参见"行列式"一章）。这也是城市空间观念发生改变的后果，根据"新住宅"运动所倡导的空间理念，追求空间的开放和流动，反对传统城市空间的封闭性。这种类型的城市空间是由自由布置的独立建筑构成的，因此带有显著的空间中的雕塑效果（图40）。

P54
功能专门化

功能、朝向和连接方式

原则上，独栋式可以和街坊式一样容纳相同的功能，尽管在大型建筑中，功能混合也是可能的（如高层住宅底部楼层用于商业），独栋式往往带有一定的功能专门化的特点，无论是特定的公共功能（市政厅、社区中心、学校、博物馆等），还是私用功能（住宅综合体、政府部门、公司总部、旅馆等）都会通过一定的建筑特征表现出来。

图40
勒·柯布西耶的马赛公寓：一种作为"垂直城市"原型的独栋式

朝向　　　　　　朝向和自然采光对独栋式来说自然不成问题。如果建筑的进深和面宽特别大，就会在室内造成暗区，除了这种情况，独栋式可以从每个面得到采光和通风。如果高层建筑布置得过于靠近，则会造成阴影问题。这种情况可以从许多城市中心区看到，如纽约以及东南亚的超大城市，包括北京、上海、香港和首尔等。

停车场　　　　　在独栋式建筑的前后左右总会有一些空地，可以设置地面停车场，然而这可能与周围空间使用的其他可能性（休闲、社会交往等）相抵触。因此，建造地下车库更为适宜。特别是对于高密度的住宅和公共建筑来说更是如此，对停车位的需求量在某些时段会比较集中。

> 提示：
> 勒·柯布西耶的"住宅是居住的机器"，特别是建于1945~1952年间能容纳1300名居民的马赛公寓尝试创造"垂直城市"，为各种收入水平的人在绿地中提供居住空间。在他出版于1935年的著作《光辉城市》中，勒·柯布西耶描绘了一种用立柱架空的多层住宅方案，将自然作为不可分离的居住空间要素。这些独栋式的居住建筑完全独立于周围的城市环境，而其内部空间却相当复杂，不但有公寓，还有商业街、交往空间、旅馆、屋顶平台和运动设施。

图41
安德里亚·帕拉第奥的圆厅别墅，维琴察

P56

历史案例

乡村环境中的个体的独立房屋、古代和中世纪的宗教设施，都可以看做基本的独栋式街区构成类型的例子。

宫殿与别墅

15世纪以来，府邸作为一种独栋式建筑在城市中的重要性日益提高，大量的金钱投入，为了表现出家族的权势和声名的显赫，这一点我们可以从佛罗伦萨的皮提宫和斯特洛奇宫以及奥格斯堡的福格尔府邸中看到。但是，到了绝对君权时期，这些附着在贵族建筑上的荣耀转向了皇家和王室建筑。其巨大尺度使城市中产阶级的略显小巧的哥特风格的住宅相形见绌。在16世纪，安德里亚·帕拉第奥在意大利维内托地区创立了一种别墅建筑样式，它的比例匀称，造型端庄精美，使之得以很快闻名于世，至今仍不失为一种典范（图41）。

城市基础设施

19世纪工业城市的扩张带来了一系列新的功能需求，诸如商业、文化、社会、政治和交通等基础设施。那些市政大厅、百货商店、剧

> 提示：
> 帕拉第奥在其著名的建筑理论著作《建筑四书》（1570年）中，将他的圆厅别墅描述成一种乡村建筑的典型，但也没打算把它置于与城市对立的位置。

113

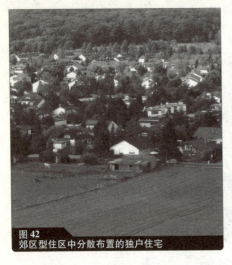
图42
郊区型住区中分散布置的独户住宅

图43
独栋式高层住宅小区的垂直密度

场、博物馆、教育机构、医院、议会建筑、火车站以及许多其他类型的公共建筑纷纷采用独栋式,使之在城市中获得了前所未有的重要地位,成为城市空间中的控制性元素。

独户住宅

汽车成为20世纪城市个人交通工具,给城市发展带来了质和量的飞跃。在一些工业化的发达国家,汽车提高了城市的机动性水平,这导致了城镇与外围地区大量独户住宅的建造热潮,进而也导致了大量开放空间的消失和乡村景观的破坏(图42)。更有甚者,这种发展模式引发了对诸如道路、排污系统等城市基础设施的大量消耗,这意味着大量的人口不得不经过很远的路程才能接近社会、文化、商业设施。

公寓式住宅

与此相反,高层建筑作为独栋式,其密度要高很多。即使如此,至少对于公寓式住宅而言,高层独栋式的建筑密度并不比4～6层的联排和街坊式高多少,因为建筑之间要留出足够的空间。还有许多独栋式住宅缺乏一种城市规划的文脉,也就是说,没有形成令人有方位感和安全感的城市空间(图43)。因此,在高层住宅的设计中,不但有必要选择适宜的建造地点,更有必要准确地定位客户群。虽然这种住宅类型并不适合带孩子的家庭、老年人和社会弱势群体,它却可以为上班族、青年夫妇、单身汉以及社会较富有阶层提供一种有吸引力的选择。

图44
法兰克福的城市天际线

城市别墅　　　近几十年建造的4~6层的城市别墅也是属于独栋式的。在城市规划术语中，城市别墅是一种介于独户住宅和高层住宅之间的形态。

摩天楼　　　摩天楼是一种特殊类型的独栋式，诸如银行大厦、公司总部等皆有意设计得很招摇以显示强有力的企业形象。然而，这些建筑从远处看的效果与在近处看大不相同。从远处看，这些摩天楼可能高耸入云，激动人心，它们甚至可以视为组团式（参见"组团式"一章），对城市轮廓产生重要影响，如法兰克福和纽约城的天际线（图44），都给人留下了深刻印象。然而，从行人和汽车的视角来看，同样的建筑物更容易被看做限定街道空间的物体而不是独栋式。因此，对其立面的设计，以及对其下部的几层室内外空间关系的处理就变得非常重要。在可能的条件下，底层部分应当设置允许公众进入的功能，这样才能为单一功能办公街区里的街道空间注入活力。

提示：
　　20世纪60年代，独栋式的作用在城市设计领域得到了重新认识。在《城市意象》一书中，凯文·林奇将那些纪念性建筑称为"标志物"，并指出这类建筑在城市空间结构的认知和定向中具有重要作用。在这部著作中，他将我们所熟悉的城市空间和建筑在脑海中的印象称为"心智地图"，这些地图虽然没有比例，却记录了个人对城市空间的体验。

图45
弗兰克·盖里设计的毕尔巴鄂古根海姆美术馆

"易位的"独栋式

近年来，随着社会的全面媒体化和全球化，出现了一种新的可以被称为"易位的"（translocational）独栋式的建筑类型。一个著名的例子是弗兰克·盖里设计的毕尔巴鄂的古根海姆美术馆，许多人并非亲眼所见，却早已烂熟于心（图45）。这座美术馆已经渗入人的意识之中，即使没见到实物，也会建立一种虚拟建筑。毫无疑问，一些早期建筑，如罗马斗兽场、比萨斜塔和埃菲尔铁塔都有类似的作用。然而，这类建筑的重要性之所以近来有大幅度增加，是由于它们在媒体上频繁露面（电视、广告等）。虽然这些建筑因尺度巨大而引人注目，它们形式上的表现力亦有过人之处。在第一个实例中，毕尔巴鄂古根海姆美术馆的建造使这座城市的访客人数大为增加。

P61

组团式

建筑成组成团布置，其特点是更多地依赖内部构成逻辑，而不是外部城市组织。高度密集且功能复合的组团也可称之为簇群。

P61

形态与空间结构

在一个组团中的每个构成元素都要与其他元素取得关联，每个元素都不能孤立看待（图46）。组团式的组织原则是：整体的构成是以相

图46
一个组团的概念图解

关联的个体为基础的。组团不可以像联排式那样进行累加（见"联排式"一章），也不可以任意扩展。

多种空间形态

组团的空间组合类型是非常有限的，但是组团式可组合的建筑类型却可以是相当宽泛的。它可以包含本书中所提到的各种类型（独栋式、行列式、联排式、庭院式和街坊式），在形态与空间上形成张力。在这些空间外形中，间距的大小、布局的疏密都是要考虑的重要因素。构成组团的建筑形态可以是开敞的，也可以是封闭的，并且往往围绕一个共同的中心、一块开放空间、一个广场、一片绿地或者由上述空间形成的一个序列来布置。正是这些空间为组团的识别性的建立发挥着特别的作用。

P62
与城市文脉脱离

城市空间构成

组团所共有的可识别性特征使之或多或少地与周围的城市文脉产生了脱离。而且，组团形成的自己的内部城市空间，由于组团的尺度和规模而呈现出不同程度的个性。

一方面，组团的个性化和可识别性可以使居民辨识自己的居住与生活环境；另一方面也存在一种危险，即不同的需求，不同的收入阶层，不同的居住人群，形成一座座"孤岛"，从而对城市社会——空间的连续性造成损害。世界各地大城市中越来越流行的封闭社区已经表明了这种危险的存在，而且后果严重。

117

> 提示：
> 封闭社区是指那种有大门且有门卫把守的居住区。进入社区要经过严格的控制，以防止诸如街头犯罪以及抢劫等来自城市环境中的危险的发生。显然，这种排他性的居住社区会造成社会隔离，但这一点正是居民所希望的。封闭社区的概念也可以用来比喻那些希望从环境中得到庇护的社会和经济团体。

P62

功能、朝向和连接方式

在很多情况下，组团式都是由居住建筑组成的，但是，也可以有其他功能。例如大学（自给自足的校园可以形成"城中城"）、医院和商务园区等。功能混合的情况也是可能的，但属于例外，因为组团式往往不能很好地和周围的城市环境相融合。因此，组团式若要混入其他功能，则需要达到一定的规模才能够维持下去。

组团式的一大优点是可以达到较高的建筑密度，而且越复杂的组团越是如此。然而，这也会造成朝向和采光方面的问题，而且当居住单元相互距离太近时，住户的私密性会受到影响。

由于中介空间和内部空间对组团的认同性至关重要，该区域中汽车交通往往不允许使用或限制使用。这样一来，就在组团中部形成了有魅力的休闲、交往空间，一个只允许行人和自行车进入的领域。停车空间可以布置在该组团边缘的地面上，或者设置多层车库，或者在交往空间之下建地下停车库。

P63

历史案例

从城市规划的角度来看，公元前2000年建立于克里特岛的弥诺斯宫和克诺索斯宫都可以被看做组团式（图47）。其复杂的空间序列和高密度形成了类似簇群一样的结构。室内就像令人晕头转向的迷宫，为"阿里阿德涅之线"的神话提供了素材（阿里阿德涅是希腊神话中克里特王弥诺斯的女儿。传说雅典每年必须进贡七对童男童女，作为关在迷宫中的牛首人身怪物弥诺陶罗斯的食物。忒修斯作为贡品来到克里特，阿里阿德涅对忒修斯一见钟情，给他一个线球，叫他捏住线头进入迷宫，杀死弥诺陶罗斯后循线而返。她随忒修斯逃离克里特，被遗弃在纳克索斯岛，在那里成了酒神狄奥尼索斯的祭司和妻子。后来，"阿里阿德涅之线"常用来表示脱出困境的方法——译者注）。阿拉伯——伊斯兰世界的城市中盘根错节的居住区也可以被看做组团式或簇群式。

图 47
克里特岛克诺索斯宫的布局

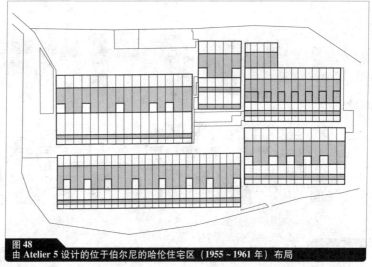

图 48
由 Atelier 5 设计的位于伯尔尼的哈伦住宅区（1955～1961 年）布局

然而，组团式主要还是近代城市发展史上的产物，它的出现主要是出于社会交往的考虑。我们可以在 19 世纪末建造的工人住宅区以及花园城市运动的设计中看到组团式的做法。如今，这种做法常常作为节省造价、节约空间的可行方法，而且通常与环境友好型、社区参与型建造活动联系在一起。组团和簇群往往采用集群设计、集群施工的方式建造。因此，组团的施工并不仅仅为了降低造价，同时也为共存提供一种建筑的、城市的表达。当然，也有一些项目是私人投资的供出售的共管住宅和独户住宅。

瑞士建筑师事务所 Atelier 5 的作品可以看做组团和簇群式建筑的上品。该事务所多年来把住宅作为自给自足的居住单元，形成了特有的风格。最早的例子是建于 1955～1961 年间的伯尔尼哈伦住宅区（图 48），而日本建筑师坂本一成于 2006 年在慕尼黑设计的制造联盟（Werkbund）住宅小区是新的例子，该设计是由高低错落的体块和私密程度不同的空地紧密地拼缀而成的（图 49）。

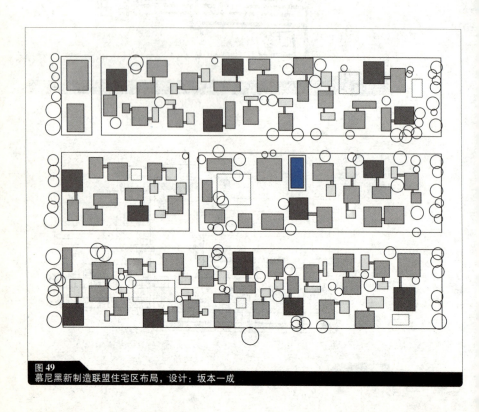

图 49
慕尼黑新制造联盟住宅区布局，设计：坂本一成

棚厦式

棚厦式是一种和独栋式相似的街区构成类型，尺度规模可大可小。这种建筑是当代城市中特有的现象，它无意与城市建立任何空间上的和文脉上的关联性。"棚厦"作为一个建筑学概念是由罗伯特·文丘里、丹尼斯·司各特·布朗和斯蒂文·伊岑诺尔在其1972年出版的著作《向拉斯韦加斯学习》中提出的。

在上述所有的街区构成类型中，棚厦式的与众不同之处在于它对外部设计的放弃。因此，它表现出强烈的反城市特征，因为它有意忽视城市的公共空间。长久以来，它根本没有被当作一种城市街区的构成类型看待，而仅仅被当作工业建筑和商业建筑不被重视。

然而，棚厦式如今变成了兴趣的焦点，其原因有二：其一，它具有开放性、灵活性，适应各种不同功能，建造经济，结构简单；其二，它的设计（或曰"非设计"）对城市及其周围的广大地区有很大影响，进而也影响到大众的日常生活空间。

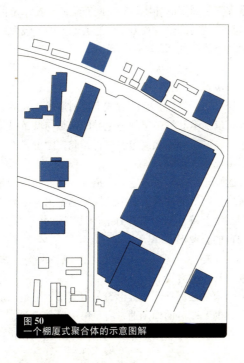

图50
一个棚厦式聚合体的示意图解

图 51
当代城市景观中的棚厦

P67　　　　　**形态与空间结构**
　　原则上说，棚厦可以设计成任何形态，只要结构上、技术上和经济上可行。它在形态、尺度上也是灵活的，可以小到一个车间，大到一个购物中心（图50）。然而，棚厦的最特别之处是其外部设计的忽视，其后果是背离了周围的环境。其空间布置源于技术上的需求以及室内空间组织与设计的需要，致力于营造有魅力的、吸引顾客的内部环境，而不把精力放在外部（图51）。

P67　　　　　**城市空间构成**
　　棚厦式随处可建。然而，它会对城市空间造成很大的破坏，因为它从根本上否定以街道和城市空间为主导的建筑与城市设计理念。因此，棚厦式应建在城市外围和周边。尽管如此，棚厦式都会对城市居民的日常生活造成很大影响。
　　外部设计的缺乏可以通过布置大型招牌、广告牌进行弥补。这些招牌和广告可以把顾客的注意力吸引到棚厦的室内（图52）。在某些情况下，通过用某种有特色的图案和分散的建筑元素来覆盖建筑立面，会形成某种独特的企业形象。

P68　　　　　**功能、朝向与连接方式**
功能的　　棚厦式可以容纳任何功能，其形态与空间结构正是功能的开放性
开放性　和可变性所导致的。棚厦的周围通常有足够的开放空间，可以布置大型的停车场。在客流较大、用地紧张的情况下，比如在市中心，可用

图52
用于昭示棚厦内部功能的广告牌和店名招牌

多层车库或地下车库来解决停车问题（图53）。

展示城市　　在棚厦内部，其外观往往戏剧性地不同于外部。在购物中心中，棚厦的室内往往设计成城市中心的气氛和效果。较早的例子是美因—陶努斯购物中心，它是20世纪60年代建于法兰克福附近的欧洲第一座购物中心，可以视为这类建筑综合体的代表。它以一条内部拱廊为中轴，两侧排列着店铺和橱窗，邀请参观者进来闲逛和逗留（图54）。为了强化城市感，一些区域被加宽，以模拟城市广场的效果，而且带有古香古色的喷泉和雕塑。购物者也可以到餐馆和冰激凌店内小憩。因此，人们从建筑内部完全察觉不到棚厦式外观缺乏设计、与周围城市环境相脱离的问题。

P69　　**历史案例**

　　本书意义上的棚厦大量出现，最早是在第二次世界大战之后。在此之前，这类建筑仅仅用于工业、交通设施和百货商店，其建筑与结构上的重要性已引起重视，著名案例包括第一次世界大战前由彼得·贝伦斯设计的位于柏林的AEG厂房，由奥古斯特·佩雷设计的位于巴黎旁蒂约路的车库（1905年）以及由伯恩哈特·策埃林设计的位于柏林莱比锡大街的蒂茨百货商店。

图53
用做停车设施的棚厦式空间

图54
棚厦内部带有展示橱窗的购物街

忽视外立面的设计

  经济上具有的优势（在某种程度上说导致了功能主义建筑普遍性的单调乏味）导致了只关注面向迎客一侧建筑外表皮设计的做法。入口以及内部的公共区域的设计具有表现性的功能，而外立面因为没有什么表现性作用而被忽视。如今，工业厂房、多层车库、品牌折扣店和购物中心的设计往往遵循这样的逻辑。

拉斯韦加斯

  在拉斯韦加斯的带状大街（Strip）上可以找到此类棚厦式的最好的例子。在那里，娱乐设施、赌场、游乐场和旅馆都采用棚厦式。灯箱和标牌反映着建筑内部功能，使建筑引人注目，且几乎代替了建筑设计和立面设计。罗伯特·文丘里、丹尼斯·司各特·布朗和斯蒂文·伊岑诺尔在他们对拉斯韦加斯的研究中，把这些结构物称为"有装饰的棚厦"。

  这种建筑在世界各地都能够看到。由于这些棚厦的影响，城市边缘特别是外围地区的景观发生了剧烈的改变。由于许多日常活动特别是那些与购物和消费相关的活动，把人们带到了这些地方，这一变化影响了生活的许多方面，也对城市和公共空间造成了影响。

虚拟世界

  这些变化随着远程通信技术的发展以及虚拟世界重要性的提升而得以加强。建筑师伯纳德·屈米指出，互联网将决定性地改变设计和我们城市的面貌，他举例说自己自从有了网上结算就再也没有进过银行大门。购物和与管理部门打交道也有类似的影响。那种从外部就能看出建筑使用功能的传统理念已经过时，一座银行没必要看起来像个银行，如果不再有人走进去，也许把它放进一个棚厦中就足矣。

结语：从城市街区到城市结构

> "建造城市，就是用三维的材料设计建筑的群体和空间。"
> ——阿尔伯特·布林克曼（1908年）

一旦每一种城市街区构成类型的特征和结构得以确认，就可以将这种认识应用到更大的（城市的）空间文脉。只有在一个城市邻里或整个城市的尺度水平上，各种建筑街区所构成的网络以及相互作用，才会构成我们每日出行、生活和工作的空间。尽管这本书分门别类地对这些街区构成的类型进行了阐述，对类型之间的相互联系的研究仍至关重要，才能理解它们对城市现实存在的理解。只有这样，才能理解城市是一个由空间的、功能的和社会的复杂系统交织而成的。

城市街区在这一复杂系统中扮演着核心的角色。城市街区作为一种物质实体，决定了个别建筑物的形态，从而也间接地影响了诸如街道、道路、广场和公园等城市中介空间。这些空间是公共（和私密）生活发生之所。由此形成的空间整体与居民和访客对于功能的使用方式，也对城市街区构成元素本身有一定影响。

不过，尽管对各种街区构成类型有个全面的了解很有必要，但是，必须把城市作为一个整体进行研究和设计。特别重要的是，面对城市在技术的、人口的、社会文化的以及经济的巨大变迁的各种挑战时，更要记住这一点。城市中有许多部分被设计建造成孤立的城市碎片，在功能层面、社会层面互不关联。那些由此而形成的"孤岛"对于某种特定功能和生活方式而言也许是比较理想的，但城市中的其他地区则会被忽视。如果不能有效地把握城市元素并对整个城市进行整合，就会引发空间和功能上的不足，并会很快给经济、社会领域带来影响。因此，把城市的不同部件整合成整体的城市结构是城市发展的核心议题。

随着21世纪的开始，规模庞大的城市群带来了城市化区域的蔓延，国际化大都市、超大城市群以及由此产生的社会分异现象对长久以来既定的观念和模式的合法性构成了挑战。不过，城市发展中的许多使命并未改变。城市发展必须创造出一个实实在在的认同感，必须创造出有特色的功能空间和社会生活空间。这其中必须包括对中介空间的设计，特别是对那些允许任何人、任何时间使用的公共空间的设计，必须寻求在公共利益和私人利益之间建立一种平衡。

因此，研究城市街区，就成了理解城市结构作为实在的生活空间和文化空间的重要性，并且致力于营造这样的空间的第一步。在此基础上，还需要与之相应的城市设计概念、程序和方法。通过对这一过程的探索，学生可以掌握一些实用的技巧和知识，这将有助于他们对城市生活空间进行概念性规划。当然，更深一步的研究以及职业实践将会引导他们超越这些基本的城市街区构成类型，去探索更为复杂的城市空间布局。只有将本书所介绍的概念放在这样一个不断发展、不断变化的背景之中，才会派上用场。

附录

参考文献

Leonardo Benevolo: *The History of the City*, MIT Press, Cambridge (Mass.) 1980

Walter Benjamin: *The Arcades Project*, Belknap Press, Cambridge (Mass.) 1999

Chuihua Judy Chung, Jeffrey Inaba, Rem Koolhaas, Sze Tsung Leong: *Project on the City 2: Harvard Design School Guide to Shopping*, Harvard Graduate School of Design 2002

Friedrich Engels: *The Condition of the Working Class in England in 1844*, J. W. Lovell Company, New York 1887

Robert Fishman: *Bourgeois Utopias. The Rise and Fall of Suburbia*, Basic Books, New York 1987

Ebenezer Howard: *Tomorrow: A Peaceful Path to Real Reform*, London 1898

Le Corbusier: *The Athens Charter*, Grossman Publishers, New York 1973

Le Corbusier: *The Radiant City*, Orion Press, New York 1967

Le Corbusier: *Towards a new Architecture*, Dover Publications, New York 1986

Kevin Lynch: *The Image of the City*, Cambridge Technology Press, Cambridge (Mass.) 1960

Franz Oswald, Peter Baccini: *Netzstadt. Designing the Urban*, Birkhäuser, Basel 2003

Andrea Palladio: *Four Books on Architecture*, MIT Press, Cambridge (Mass.) 1997

Philippe Panerai, Jean Castex, Jean-Charles Depaule: *Urban Forms*, Architectural Press, Boston 2004

Aldo Rossi: *The Architecture of the City*, MIT Press, Cambridge (Mass.) 1982

Camillo Sitte: *The Art of Building Cities. City Building according to its Artistic Fundamentals*, Reinhold Publishing Corporation, New York 1945

Raymond Unwin: *Town Planning in Practice: An Introduction to the Art of Designing Cities and Suburbs*, T. F. Unwin, London 1909

Robert Venturi, Denise Scott Brown, Steven Izenour: *Learning from Las Vegas*, MIT Press, Cambridge (Mass.) 1972

P74
作者简介
  托尔斯滕·别克林，哲学博士，工学硕士，美因河畔法兰克福应用科技大学城市规划与建筑科学学科讲师，卡尔斯鲁厄市从业建筑师。
  迈克尔·彼得莱克，工学博士，美因河畔法兰克福应用科技大学城市规划与设计系教授，美因河畔法兰克福市从业城市规划师。